ELECTROMECHANICAL COUPLING OF THE SOLAR ATMOSPHERE

AIP CONFERENCE PROCEEDINGS 267

ELECTROMECHANICAL COUPLING OF THE SOLAR ATMOSPHERE

CAPRI, ITALY 1991

EDITORS:

DANIEL S. SPICER
NASA/GODDARD SPACE FLIGHT CENTER

PETER MACNEICE
HUGHES STX

American Institute of Physics   New York

Authorization to photocopy items for internal or personal use, beyond the free copying permitted under the 1978 U.S. Copyright Law (see statement below), is granted by the American Institute of Physics for users registered with the Copyright Clearance Center (CCC) Transactional Reporting Service, provided that the base fee of $2.00 per copy is paid directly to CCC, 27 Congress St., Salem, MA 01970. For those organizations that have been granted a photocopy license by CCC, a separate system of payment has been arranged. The fee code for users of the Transactional Reporting Service is: 0094-243X/87 $2.00.

© 1992 American Institute of Physics.

Individual readers of this volume and nonprofit libraries, acting for them, are permitted to make fair use of the material in it, such as copying an article for use in teaching or research. Permission is granted to quote from this volume in scientific work with the customary acknowledgment of the source. To reprint a figure, table, or other excerpt requires the consent of one of the original authors and notification to AIP. Republication or systematic or multiple reproduction of any material in this volume is permitted only under license from AIP. Address inquiries to Series Editor, AIP Conference Proceedings, AIP, 335 East 45th Street, New York, NY 10017-3483.

L.C. Catalog Card No. 92-82717
ISBN 1-56396-110-5
DOE CONF-9105357

Printed in the United States of America.

CONTENTS

Flux Tube Sizes and Temporal Evolution 1
    T. J. Bogdan
Large Scale Simulations 13
    Å. Nordlund and K. Galsgaard
Magnetic Flux Tubes as Communication Channels 24
    B. Roberts
Magnetic Field Line Topology in Solar Active Regions 35
    N. Seehafer
Weak Solar Magnetic Fields 40
    J. O. Stenflo
Magnetic Fields, Oscillations, and Heating in The Quiet Sun
Temperature Minimum Region 55
    J. W. Cook
Explosive Events and Magnetic Reconnection in the Solar Atmosphere 63
    K. P. Dere
Three-Dimensional Kinematic Reconnection of Plasmoids with Nulls 71
    Y.-T. Lau and J. M. Finn
Generation of Magnetic Fields by Chaotic Fluid Convection:
The Fast Dynamo Problem 79
    J. M. Finn
Understanding the Source of the Solar Activity Cycle: Results and
Prospects from Helioseismology 85
    P. R. Goode
Alternative Coronal Heating Mechanisms 100
    R. N. Sudan and D. W. Longcope
Coronal Heating through Lack of MHD Equilibrium 111
    P. C. H. Martens, M. T. Sun, and S. T. Wu
On the Collective Appearance of Coronal Loops and the
Resistive Heating Instability 116
    Y.-Q. Lou
Alfvén Waves in Current-Carrying Inhomogeneous Plasmas 121
    H. Shigueoka, C. A. de Azevedo, A. S. de Assis, and P. H. Sakanaka
The X-ray Ultraviolet Imager for the Orbiting Solar Laboratory 126
    E. Antonucci, M. Malvezzi, L. Ciminiera, F. Angrilli, M. E. Bruner,
    G. Perona, M. A. Dodero, B. L. Evans, L. Golub, M. Landini,
    G. Noci, P. McWhirter, B. M. Fossi, G. Poletto, D. F. Neidig,
    W. K. H. Schmidt, R. J. Thomas, and G. Tondello
Hydrostatic Models of X-ray Coronal Loops Observed by NIXT 136
    G. Peres, F. Reale, and L. Golub
The Effect of Viscosity on Hydrodynamics of Coronal Flares 140
    F. Reale and G. Peres

On Non-local Transport Processes in the Solar Atmosphere 145
    P. MacNeice
MHD Turbulence in an Expanding Atmosphere 154
    M. Velli, R. Grappin, and A. Mangeney
Hot Mass Transport in the Solar Active Prominence 160
    A. Kučera, M. Saniga, and J. Rybák
Author Index 169

Preface

The OSL Workshop on Electromechanical Coupling of the Solar Atmosphere held on the island of Capri, Italy during the week of May 27–31, 1991, resulted from a desire by the Orbiting Solar Laboratory (OSL) Science Working Group (SWG)* to have an ongoing series of workshops designed to emphasize high-resolution solar physics. Due to the suspension of the OSL Program, the Capri meeting was, unfortunately, the last in a series of three.

It was anticipated that, in the near future, extremely high-resolution measurements of solar magnetic and velocity fields, together with densities and temperatures from the photosphere through to the corona, were expected to be made by NASA's planned OSL. Thus, the principal objective of the third OSL workshop was an in-depth assessment of our present theoretical and experimental understanding of how radiation, velocity, and magnetic fields, together with thermodynamic variables, behave as a fully coupled system. OSL measurements were expected to result in dramatic improvements in our understanding of how mass, momentum, and energy are transported from one portion of the Sun to another. These physical quantities tend to be spatially localized by magnetic fields, which are believed to play both a passive and active role in the transport processes. The role of magnetic fields in the electromechanical coupling of different parts of the solar atmosphere was therefore a key issue of the workshop.

To address these issues and to help provide strong theoretical support for the OSL mission, the workshop brought together not only solar physicists, but also experts in the allied specialties of dynamo theory, chaos theory, MHD turbulence, and wave propagation theory. This provided a forum where the issues alluded to in the previous paragraph were fully and openly discussed. The workshop split into a series of topics, some global in scope, some local, and each with an assigned topic reviewer. The reviewer's brief was to play the role of devil's advocate by providing a critical assessment of the topic under consideration and then to act as focus for the discussion period that followed. In addition, all the OSL instrument PI's gave a detailed summary of how their respective instruments were designed and of the measurements each instrument would make.

Although the OSL mission is suspended, the workshop made it abundantly clear that the need for an integrated mission such as OSL will not disappear if solar physics is to make any significant scientific progress. In particular, the interaction of ionized gases and magnetic fields—both of which are ubiquitous in the universe—which is the cause of much complex solar activity such as solar flares, cannot be understood without an OSL. In addition, since the Sun provides the only stellar plasma that is accessible for detailed observation and analysis, no real understanding of astrophysical phenomena involving magnetized plasmas will ever be complete without the measurements OSL was designed to make. Thus, OSL will eventually happen.

The editors would like to thank the members of the workshop science organizing committee for putting together a well-balanced series of topics with an excellent set of speakers.

Daniel S. Spicer
Peter MacNeice

---

*The OSL SWG consisted of Ester Antonucci, Guenter Brueckner, George Doscheck, Richard Fischer, Jack Harvey, Bruce Lites, Robert Noyes, Wolfgang Schmidt, Dan Spicer (Chairman), Alan Title, Howard Zirin, and Cornelis Zwaan.

FLUX TUBE SIZES AND TEMPORAL EVOLUTION

T. J. Bogdan
High Altitude Observatory
National Center for Atmospheric Research[*]

ABSTRACT

The present observational knowledge of the size distributions of solar surface magnetic structures—sunspots, sunspot groups and active regions—and their temporal evolution, is reviewed in the context of how such information may provide important clues to the nature of the solar dynamo and the underlying causes of solar variability. The ability of such information to distinguish between the competing theoretical mechanisms of flux tube fragmentation and coalescence is briefly discussed.

INTRODUCTION

This contribution attempts to synthesize several diverse studies of the size distributions of solar surface magnetic flux concentrations ranging in size from the sub-arcsecond elements upward to active regions covering several percent of the visible solar disk. The underlying motivation stems from both an intellectual, and more recently a climatic,[7,9] need to understand the origins of solar variability. Used in concert with the more traditional indicators of the solar cycle, knowledge of the solar cycle variation of the distribution of sizes—or equivalently, magnetic fluxes—of emerging magnetic fields is likely to provide important new clues to the basic mechanisms underlying origins of solar, and stellar[17] variability and perhaps ultimately the nature of the dynamo.

In order to be able to draw such inferences, one requires an underlying theoretical template against which the observations may be interpreted. Accordingly, this contribution also surveys the extant theoretical studies which have attempted to predict the size spectrum of emerging magnetic flux based upon various potentially relevant physical processes. Such studies are rare, because until only very recently, there has been a lack of comprehensive and accessible observational analyses. For example, the extensive theoretical modelling of the solar dynamo carried out over the last two decades attempted merely to induce the magnetic flux to emerge at the proper latitudes at the proper times, with no mention of the distribution of sizes of these emerging magnetic flux concentrations (for a notable exception, see Schüssler[19]).

One of the unexpected unifying themes that will emerge from this contribution is the ubiquitous presence of the lognormal distribution (the logarithm of a given quantity is normally distributed) in both the observational and theoretical contexts. The observed size distributions of sunspot umbrae and probably sunspot groups are both lognormal. The linear decay rates of sunspot group areas are also lognormally distributed. Theoretical studies of the chaotic gen-

---

[*] The National Center for Atmospheric Research is sponsored by the National Science Foundation

2 Flux Tube Sizes and Temporal Evolution

eration of magnetic fields by the so-called fast dynamos[8] appear to routinely produce lognormally distributed magnetic field strengths $|\mathbf{B}(\mathbf{x}_i)|$ for randomly chosen points $\mathbf{x}_i$ as a natural consequence of chaotic (non-integrable) streamlines. A connection, if any, that exists between the lognormal distributions arising in these two contexts related to the emergence of solar magnetic flux is intriguing.

The next three sections discuss the measured distributions of sunspot umbrae, sunspot groups, and active regions, respectively. The particulars of the individual data sets and corresponding references can be found in Table I. The next to last section of the paper discusses size distributions which result from the competing processes of the fragmentation of large flux regions into small flux tubes, and the coalescence of small flux elements to form larger flux tubes.

Table I Data Sets.

	Period	References	Source
Sunspot Umbrae	1917-1982	4	Mt. Wilson White-Light Plate Collection
Sunspot Groups	1874-1976	11,14,16	Greenwhich Photoheliospheric Results
Active Regions	1967-1981	21	Mt. Wilson Magnetograms
Active Regions	1975-1986	10	NSO Full-disk Magnetograms
Active Regions	1973-1980	18	Solar Geophysical Data Ca II K Plage Areas

SUNSPOTS: DISTRIBUTION OF UMBRAL AREAS

The size distribution of sunspot umbral areas has been studied by Bogdan et al[4], using data obtained from the Mt. Wilson white-light plate collection[11]. The period covered is the years 1917 to 1982, which includes cycles 15 to 21. Two selection criteria were applied to arrive at the final data set: (i) the spots had to be situated within $\pm 7.5°$ longitude and $\pm 45°$ latitude of disk center, and (ii) after correction for foreshortening, the measured umbral area had to exceed 1.5×10^{-6} $A_{\odot/2}$ (1×10^{-6} $A_{\odot/2}$ is an area corresponding to one millionth of the visible solar disk). There were 24,615 sunspots which satisfied these two criteria. Sunspots that lived longer than one solar rotation and fulfilled the above two criteria on each passage across the disk were counted as separate entities on each central meridian passage. Since individual sunspots were not followed over several successive daily white-light images, it was only possible to assign the instantaneous umbral area to a given spot when it satisfied the two criteria. The relationship between this instantaneous area and the maximal umbral area is difficult to ascertain without knowledge of the variation of area with time (see below).

The differential size distribution,

$$\frac{dN}{dA} \cdot \Delta A$$

which is the frequency of occurrence of a sunspot umbral area in a range ΔA about the value A, was found to be well-fit by a parabola on a log-log plot, indicating that the logarithms of the sunspot umbral areas are normally distributed. Moreover, the same distribution was found to obtain throughout the solar cycle and from one cycle to the next. Only the overall normalization of the lognormal distribution varied during the solar cycle reflecting the familiar cyclic variation of the sunspot number (see figure 1).

The mean and variance of the (normal) distribution of $\ln(A)$ for sunspot umbrae appear to be robust parameters that provide an important intrinsic characterization of flux generation in the solar interior. Uncertainties in their precise values result from the fact that the mean value lies somewhat below the spatial resolution of the Mt. Wilson white-light plates (see figure 1). The mean value of $\ln(A)$ is situated between -1.1 and -0.48, corresponding to umbral radii in the range of 320km and 430km (0.34-0.62×10^{-6} $A_{\odot/2}$). The variance is bounded by 1.16 and 1.25. The mean and variance values quoted above assume that the lognormal distribution continues to be valid for umbral areas below the resolution of the Mt. Wilson plates. Under the same assumption, the mean of A over the lognormal distribution lies between 1.0-1.3×10^{-6} $A_{\odot/2}$.

SUNSPOT GROUPS: AREAS AND LIFETIMES

Kopecký[12] provides a plot of the differential distribution of the areas of sunspot groups in a format similar to that of figure 1. He uses data from the Greenwich Photoheliospheric Results covering the period from 1913 to 1954 and including cycles 15-18. The selection criterion is that the sunspot group must lie within ±30° longitude of disk center. The plotted data cover the range 1-1500×10^{-6} $A_{\odot/2}$, but no attempt is made to fit these data. An examination of the figure shows that a simple linear regression is unacceptable for a plot of log frequency of occurrence versus log area; it appears that a parabolic fit very similar to that of figure 1 may be sufficient. Kopecký finds that, up to an overall normalization, the distribution of sunspot group areas appears to be nearly the same for each of the four cycles. The mode of the distribution lies between 3-5×10^{-6} $A_{\odot/2}$. No discussion of the reliability or uncertainties of the data is given.

Moreno-Insertis and Vázquez[16] have also made use of the Greenwich Photoheliospheric Results to study the decay rates of sunspot groups. They used data from the period 1874-1939 spanning cycles 11-17. They imposed three selection criteria: (i) the sunspot group had to be situated within ±60° of the central meridian, (ii) the area had to exceed 35×10^{-6} $A_{\odot/2}$ for the measured umbra and penumbra areas, and 15×10^{-6} $A_{\odot/2}$ for the umbral area alone, and (iii) the group area had to be measured at five or more distinct times. Some 1515 sunspot groups were found to satisfy these constraints. Of these, roughly 5% were found to decay exponentially with time and were subsequently discarded.

4 Flux Tube Sizes and Temporal Evolution

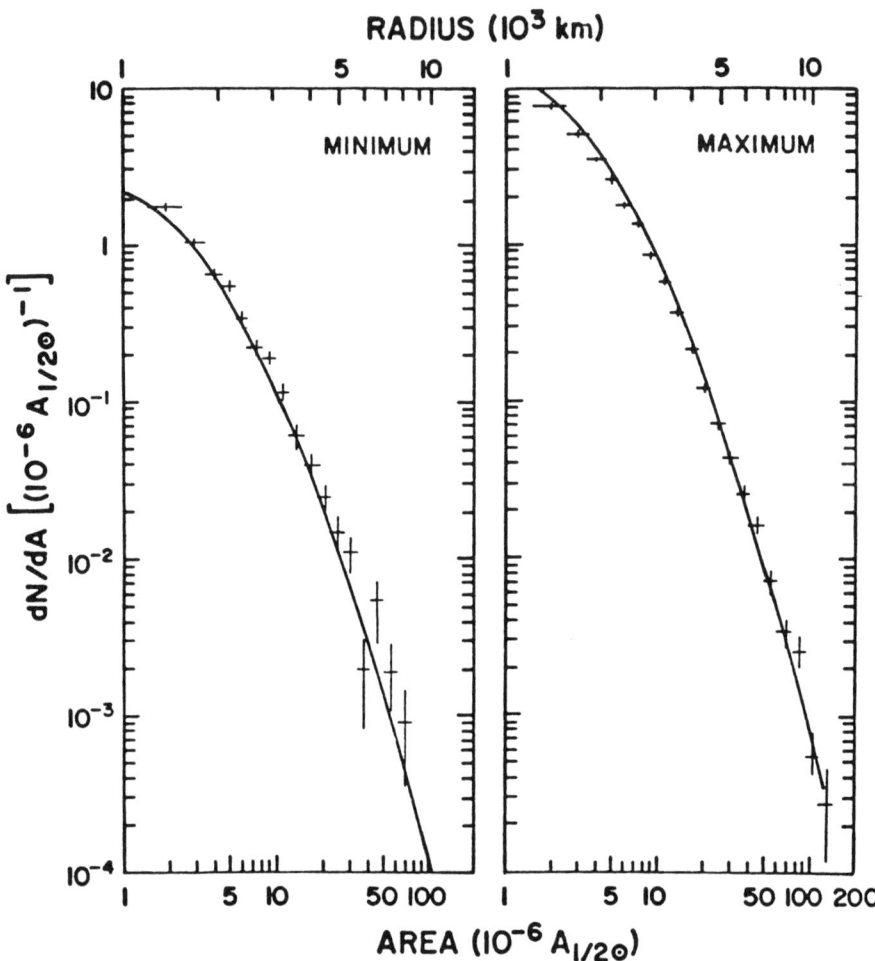

Fig. 1. Differential distribution of sunspot umbral areas at solar maximum and solar minimum. The same lognormal distribution (solid line) has been used to fit both sets of data using only a different overall normalization [from Bogdan et al[4]].

For the remaining 1436 groups, a decay law of the form:

$$A(t) = A_0 \left[1 - (0.9 + q)\frac{t}{t_\star} + q \left(\frac{t}{t_\star}\right)^2 \right] \qquad (1)$$

was fit to the data. The quantities A_0 and $t_\star$ are the characteristic area and decay time for the group, while the quantity q measures any departure from a constant rate of decay. The mean value of q was found to be essentially zero for the

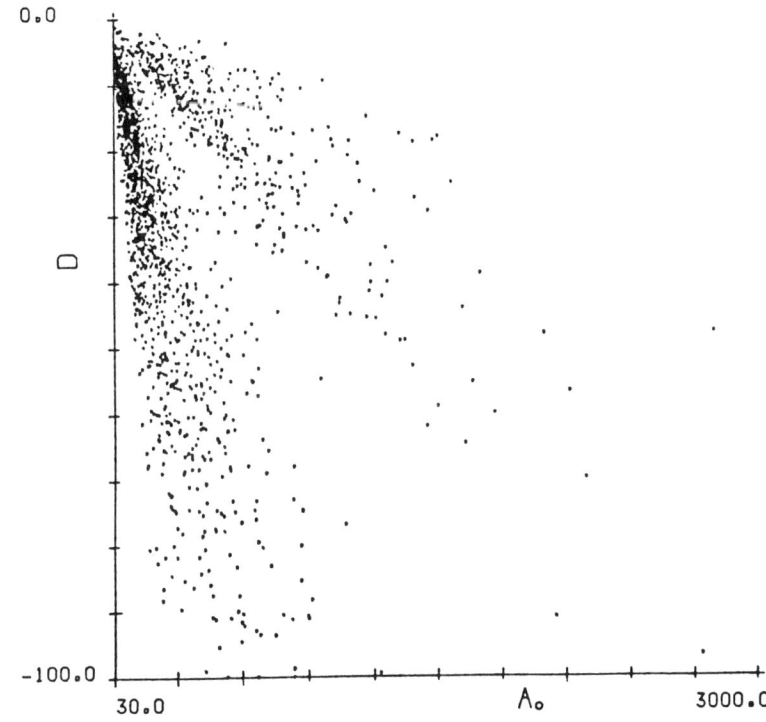

Fig. 2. Linear decay rate for sunspot groups (measured in 10^{-6} $A_{\odot/2}$ day^{-1}) versus maximum area of the sunspot group (umbra and penumbra measured in 10^{-6} $A_{\odot/2}$) for the entire sample of 1436 groups [from Moreno-Insertis and Vázquez[16]].

entire data set as well as the various subsets—complex spot groups (LaLaguna Type 2), groups containing isolated spots (LaLaguna Type 3), recurrent and non-recurrent groups—indicating that on average the area of a sunspot group decreases linearly with time. In this case, the characteristic time t_* is simply related to the decay rate D according to $t_* = 0.9 A_0/D$, while A_0 is the maximum area of the group.

Equation (1) was then used once more to fit the sunspot group areas but with q set equal to zero. Figure 2 shows the dependence of D upon A_0. Contrary to popular opinion, there is clearly no obvious relationship between the decay rate and the area of the sunspot group. There is a triangular area that is noticeably underpopulated, but this is an artifact of the selection criteria (i) and corresponds to characteristic decay times between 15 and 22 days. A trend similar to that shown in figure 2 was found for each of the subsets. The lack of any clear correlation between the maximum area and the decay rate for sunspots groups suggests that the use of the instantaneous area for the sunspot group area in lieu of the maximum area may be relatively free from area-dependent selection biases.

Martínez Pillet et al[14] have studied the distribution of linear decay rates for sunspot groups employing the Greenwich data from a somewhat longer period 1874-1976, including cycles 11-20. Surprisingly, the frequency of occurrence of the decay rates appears to be lognormally distributed. This is illustrated by figure 3 which shows the same distribution plotted side by side in both linear-linear and log-log formats.

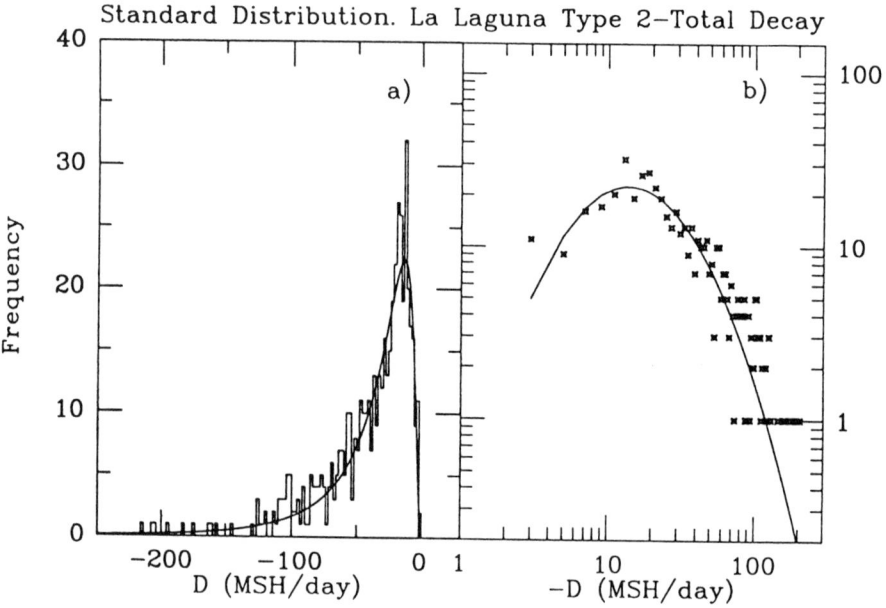

Fig. 3. Frequency of occurrence of the decay rate for sunspot groups (umbra and penumbra) for complex spot groups from cycles 15-17 and 19. The total sample contains 515 decay rates [courtesy of V. Martínez Pillet].

ACTIVE REGIONS: AREA DISTRIBUTION

Tang et al[21] carried out the first study of the size distribution of active regions employing Mt. Wilson daily magnetograms from the period 1967-1981. They defined an active region as a bipolar magnetic structure consisting of features at the level of 10 Gauss or more. Each region was treated as a separate entity on each of its disk passages. The total data set consisted of 5086 active regions with areas between 3 and 1350 deg^2 (1 deg^2 = 48.5×10^{-6} $A_{\odot/2}$). Figure 4 shows the cumulative distribution [$N(A)$ is the frequency of occurrence for all active regions with areas in excess of A] for various subsets of the data plotted in a log-linear format. At solar maximum the active region areas are apparently distributed exponentially, while at solar minimum (middle panel) there appears to be an excess of small active regions.

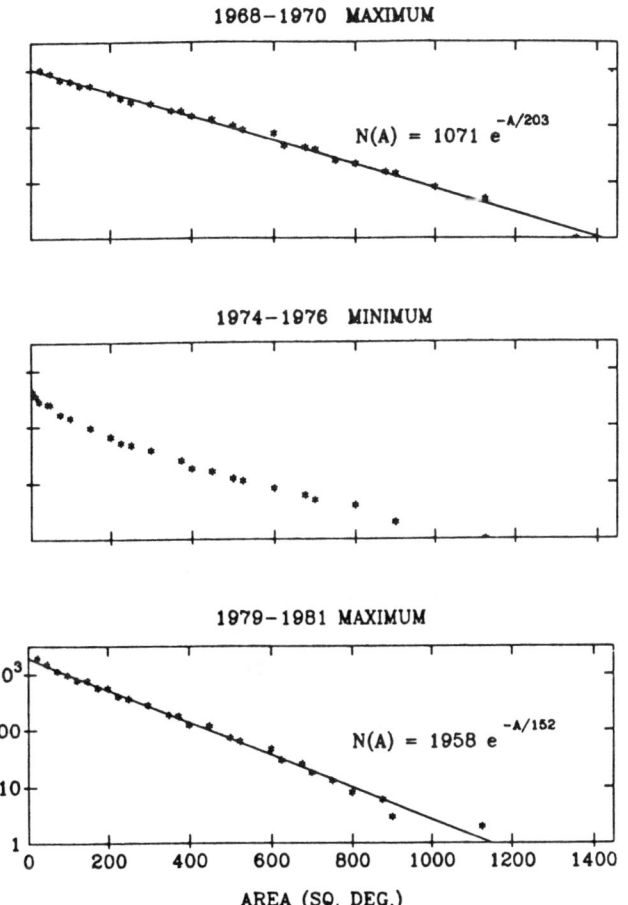

Fig. 4. Cumulative distribution of large active region areas for solar maximum and solar minimum periods [from Tang et al[21]].

Harvey[10] has carried out a complementary study using National Solar Observatory daily full-disk magnetograms from the period 1975-1986, and focusing on smaller active regions. Some 978 active regions comprise her data set. Unlike the study of Tang et al, the active regions are not counted again on their subsequent disk passages. In addition, careful attention was paid to the temporal development of the active regions: the active regions were required to pass through their maximum development on the visible solar disk, and the size distribution for the sample was determined using this maximum area. The resulting cumulative distribution is shown in figure 5. The cumulative distribution is not too inconsistent with a simple power-law, although a cubic dependence of ln(N) upon ln(A) is necessary if all the points are to be fit closely. The bottom panel of figure 5 shows clearly that the distribution is independent of cycle phase up to an overall normalization, reflecting the fact that there are 5-6 times more active regions present at solar maximum than at solar minimum.

The smaller active regions studied by Harvey appear to behave quite differently than those studied by Tang et al. In part this may be a consequence of the different methods of analysis employed in each study. A third study of size distributions of active regions was carried out by Schrijver[18]. He used Ca II K plage area listed in the Solar Geophysical Data. His plot of the differential area distribution on a log-log plot shows curvature similar to that of the sunspot umbral areas shown in figure 1 above. The area included in his sample range from 100-7000×10^{-6} $A_{\odot/2}$. Nearly the same distribution is obtained throughout the solar cycle, although similar to the results of Tang et al, an excess of small active regions ($A \leq 200 \times 10^{-6}$ $A_{\odot/2}$) appears at solar minimum. Schrijver chose to fit the data with two different exponential profiles, one valid for areas below 1660×10^{-6} $A_{\odot/2}$, and the other valid for larger areas. It is evident from the figure in Schrijver's paper that this particular three parameter fit results in a curve very similar to the lognormal distribution in the range of areas of interest.

FLUX TUBE COALESCENCE vs. FRAGMENTATION

The size spectrum of magnetic flux tubes arising from the coalescence of small flux tubes into larger flux tubes has been considered in a series of papers by Bogdan and Lerche[2,3,5,6]. The essential features of such a process are illustrated simply by the following model. Suppose there is a concentration $C_1(0)$ at time $t = 0$ of identical magnetic flux tubes of area $A_\star$ (or equivalently flux $\Phi_\star$). Through binary collisions these flux tubes coalesce to form flux tubes of area $2A_\star$, the concentration of these two-tuples increasing at the rate $\sigma_{1,1} C_1^2(t)$. These two-tuples can then coalesce with the original population to form three-tuples, and can coalesce among themselves to form four-tuples. The governing equations for the concentration of k-tuples, $k = 1, 2, 3, ...$, are easily shown to be:

$$\frac{dC_k}{dt} = \frac{1}{2} \sum_{i+j=k} \sigma_{i,j} C_i C_j - C_k \sum_{j=1}^{\infty} \sigma_{k,j} C_j , \qquad (2)$$

which are known as the stochastic collection equations. A remarkable property of equation (2) is that it possess an exact analytic solution for the case in which the coalescence process is independent of the size of the flux tubes: $\sigma_{i,j} = \sigma$ independent of i and j. For an initial concentration $C_1(0)$ of single flux tubes one finds,

$$C_k(t) = C_1(0) \frac{4}{(2+\tau)^2} \left(\frac{\tau}{2+\tau} \right)^{k-1} , \qquad (3)$$

where $\tau = t\sigma C_1(0)$. Since the area of a given k-tuple is simply $kA_\star$, equation

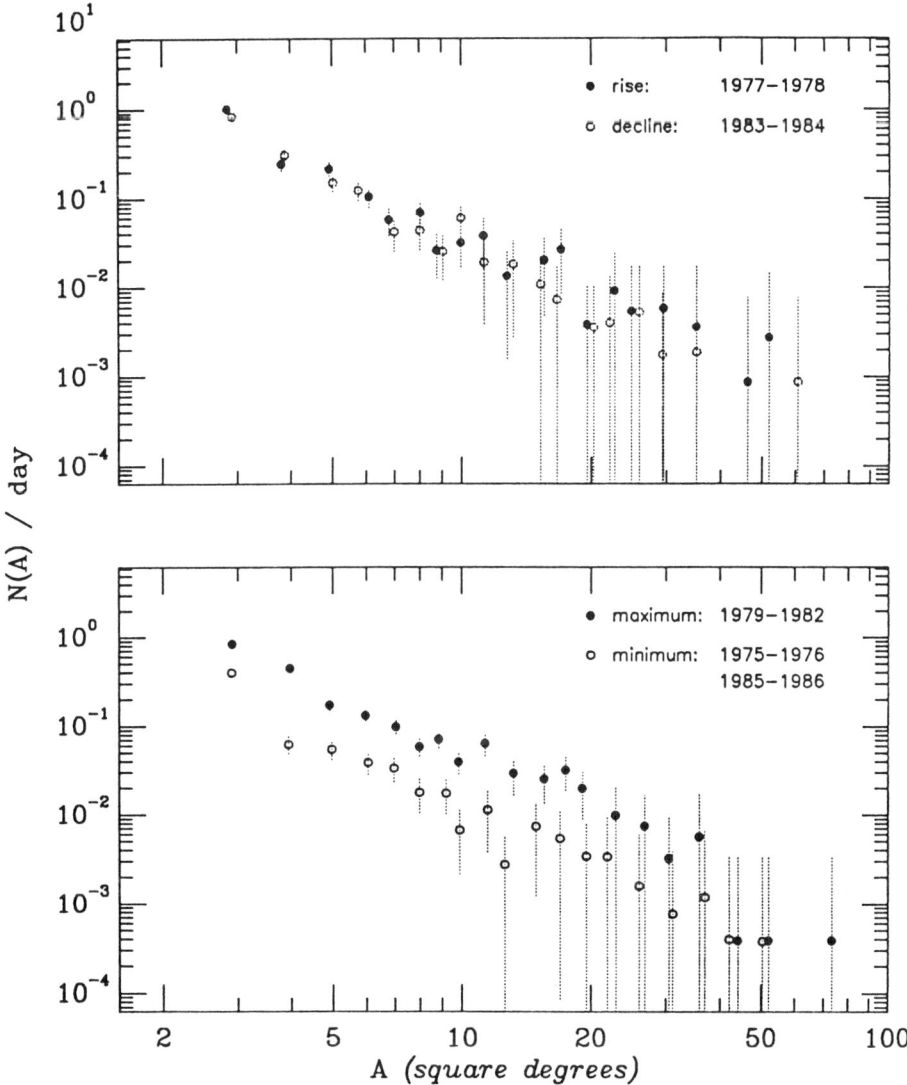

Fig. 5. Cumulative distribution of small active region areas for different phases of the solar cycle [from Harvey[10]].

(3) predicts that the differential size distribution of flux tubes formed through such a coalescence process is an exponential distribution $\exp(-A/A_c)$, with a characteristic area scale $A_c = A_\star/\ln(1 + 2/\tau)$, where τ depends upon the initial concentration of flux tubes, the elapsed time, and the coalescence rate. Such a distribution is consistent with the active region size distribution at solar maxi-

mum found by Tang et al. Of course the shape of the distribution is dependent upon the choice of the coalescence rate, and the proper choice can only be determined from a detailed treatment of the binary interaction of magnetic flux tubes. A rudimentary discussion of such collisions between twisted magnetic flux tubes has been given by Bogdan[2]. It should be noted that atmospheric aerosols formed through coagulation exhibit nearly lognormal size distributions when a realistic $\sigma_{i,j}$ is employed in equation (2) above.[13]

Fragmentation is the antithesis of coalescence. The archetypal example of the fragmentation process views the size distribution of flux tubes, for example, as the n-th generation of a binary tree formed by a branching process. At step 0 there is one large magnetic flux rope of area $A_\star$, the ancestor. This flux tube then fragments into a pair of offspring of areas $xA_\star$ and $(1-x)A_\star$ which are the first generation. The quantity x is a random variable assumed to lie between 0 and 1, and $p(x)$ is the probability distribution of choosing a certain x, with

$$p(0.5 - x) = p(0.5 + x)$$

and

$$\int_0^1 dx\, p(x) = 1.$$

Each of these two offspring in turn fragment into two more flux tubes providing the second generation, and so on. For the present example, each time a flux tube fragments into a pair of flux tubes the random variable x describing this process is determined by the same probability distribution $p(x)$. When the n-th generation is reached there are 2^n flux tubes with areas $A_1, A_2, A_3, ..., A_{2n}$, any one of these areas can be written as $\prod_{i=1}^{n} x_i A_\star$—the product of n random variables. The logarithm of the any one of these areas is therefore the sum of n logarithms of random variables, and hence in the limit of large n, the central limit theorem can be invoked to show that the distribution of $\ln(A_1/A_\star)$, $\ln(A_2/A_\star), \ln(A_3/A_\star), ..., \ln(A_{2n}/A_\star)$ is normal.[1,15] The mean

$$m = (n+1) \int_0^1 dx\, \ln(1-x)\, p(x) \tag{4}$$

and the variance

$$\sigma^2 = -\frac{m^2}{n+1} + (n+1) \int_0^1 dx\, \ln^2(1-x)\, p(x) \tag{5}$$

depend upon $p(x)$, but the lognormal nature of the distribution is independent of the particular choice for $p(x)$, provided the same $p(x)$ is used to create each successive generation.

This distribution formed through flux tube fragmentation is consistent with the observed size distribution of sunspot umbral areas (compare above with large active regions). If one naively applies equations (4) and (5) to the distribution of sunspot umbral areas, then one is led to conclude that $A_\star \gg 10^{-6} A_{\odot/2}$ and hence $n \gg 1$. However since σ^2 is of order unity, $p(x)$ must be highly peaked about $x = 0.5$, suggesting that flux tubes should typically fragment into two offspring of nearly equal magnetic flux.

CONCLUSIONS AND OUTLOOK

In conclusion, it is fair to say that more questions have been raised than answers provided. Despite the bewildering array of facts that are accumulating on the size distribution of solar surface magnetic flux concentrations, a few important unifying themes are present. The most striking is that the size distributions of sunspot umbrae and small active regions are independent of solar cycle, and then even for large active regions and Ca II K plage where cycle variations have been noted, these variations are not particularly striking and reside exclusively in the smallest sized objects in the data set. This strongly suggests that the means by which the Sun produces the size spectrum of magnetic structures is not modulated by the numbers of such structures produced. This point argues against an important role for the coalescence process in the formation of magnetic structures because the characteristic scale $A_c = A_\star/\ln(1 + 2/t\sigma C_1(0))$ depends sensitively on the concentration of small flux tubes, and hence compensating variations of the elapsed time t and/or the coalescence rate σ must occur for the resulting size distributions to be nearly independent of the solar cycle (flux tube concentration).

Neither is the fragmentation origin on particularly solid ground. There is no available evidence to support the notion that active regions are lognormally distributed, particularly for the largest sized active regions (see figure 4). This may result from the possibility that active regions, containing the largest amounts of unsigned flux, are not far removed from the ancestor in the branching process, and hence the distribution is not lognormal as it should be in the limit of large n. The distribution of small active regions (see figure 5) shows some tendency for a parabolic lognormal profile, but then an excess of the smallest active regions vitiates that trend for $A < 6$ deg^2. Sunspot groups appear to be lognormally distributed over some range of areas, but issues of spatial resolution and visibility raise serious concerns.

It is imperative to try to measure the size distribution for the smallest magnetic flux concentrations. Spruit and Zwaan[20] have made an important first step in this direction, in their attempt to infer the size distributions of pores knots and elements. The difficulties encountered here are even more severe than in the studies discussed above. In addition to the obvious lack of sufficient spatial resolution the intensity contrast between the flux tube and the surrounding quiet Sun is a very strong function of the diameter of the flux tube. OSL should be particularly well-suited to overcome some of these limitations and provide important new diagnostic information on the size distributions of the smallest magnetic features.

ACKNOWLEDGMENTS

I thank K. Harvey and V. Martínez Pillet for generously making available results of their research prior to publication. I thank F. Moreno-Insertis and F. Tang for providing figures from their papers in a timely fashion, and J. Sánchez Almeida for his comments on the manuscript.

REFERENCES

1. J. Aitchison and J.A.C. Brown, The Lognormal Distribution (Cambridge Univ. Press, 1973).
2. T.J. Bogdan, Phys. Fluids 27(4), 994 (1984).
3. T.J. Bogdan, Astrophys. J. 299, 510 (1985).
4. T.J. Bogdan, P.A. Gilman, I. Lerche and R. Howard, Astrophys. J. 327, 451 (1988).
5. T.J. Bogdan and I. Lerche, Astrophys. J. 296, 719 (1985).
6. T.J. Bogdan and I. Lerche, Physica 25D, 382 (1987).
7. G. Brasseur, A. De Rudder, G.M. Keating and M.C. Pitts, J. Geophys. Res. 92D1, 903 (1987).
8. J.M. Finn and E. Ott, Phys. Rev. Lett. 60(9), 760 (1988).
9. P. Foukal, J.Geophys. Res. 92D1, 801 (1987).
10. K.L. Harvey, Solar Phys. , to be submitted (1992).
11. R. Howard, P.A. Gilman and P.I. Gilman, Astrophys. J. 283, 273 (1983).
12. M. Kopecký, BAC 15(2), 44 (1964).
13. K.W. Lee, H. Chen and J.A. Gieske, Aerosol Sci. & Technology 3, 53 (1984).
14. V. Martínez Pillet, F. Moreno-Insertis and M. Vázquez, Astrophys. Space Sci. 170, 3 (1990).
15. E.W. Montroll and M.F. Shlesinger, Proc. Natl. Acad. Sci. USA 79, 3380 (1982).
16. F. Moreno-Insertis and M. Vázquez, Astron. Astrophys. 205, 289 (1988).
17. R.R. Radick, G.W. Lockwood and S.L. Baliunas, Science 247, 39 (1990).
18. C.J. Schrijver, Astron. Astrophys. 189, 163 (1988).
19. M. Schüssler, Nature 288, 150 (1980).
20. H.C. Spruit and C. Zwaan, Solar Phys. 70, 207 (1981).
21. F. Tang, R. Howard and J.M. Adkins, Solar Phys. 91, 75 (1984).

LARGE SCALE SIMULATIONS

Åke Nordlund and Klaus Galsgaard
Copenhagen University Observatory

ABSTRACT

We discuss large scale numerical simulations as a tool for obtaining qualitative understanding of the processes directly and indirectly responsible for coronal heating. The actual heating process in the low beta coronal plasma is most likely driven by transfer of magnetic energy from the subsurface high beta region, where magnetic energy is created as an energetically insignificant byproduct of solar convection and rotation. Based on the results of recent numerical experiments, we discuss some of the processes involved.

1. INTRODUCTION

There is broad consensus, in general terms, on the chain of mechanisms that is ultimately responsible for the existence of a hot solar corona. The solar energy flux is carried through the outer 30% of the solar radius by convection, with motions on a large range of scales. A tiny fraction of the mechanical energy associated with the convective motions is, under the influence of (differential) rotation, converted to magnetic energy. The major part of this pool of magnetic energy resides in the convection zone itself. A small fraction penetrates the solar surface and extends into the solar corona but is still anchored in the convection zone. A flux of magnetic energy (Poynting flux) is associated with the motion of the footpoints of the coronal field, and with the continual emergence of new field through the solar surface. The Poynting flux tends to increase the free (non-potential) energy of the coronal magnetic field and conversion of magnetic energy into thermal or kinetic energy tends to decrease the free magnetic energy of the corona. A quasi-static energy balance is achieved when the coronal magnetic field dissipates as much energy on the average as it receives through the solar surface.

As long as the discussion is kept in these general terms, there can be little or no doubt about the validity of the scenario. It is the detailed mechanisms in each link of the chain that need to be clarified and understood. For example, is most of the Poynting flux associated with rapid wave-like (AC) motions, or is most of it associated with slower systematic (DC) motions? What are the actual mechanisms in the corona that, on the average, dissipate enough magnetic energy to balance the Poynting flux through the solar surface? Empirically, we know that the balance is not a detailed one. There are large scale catastrophic flare events that dissipate a lot of energy in a short period of time. Is this peculiar for the large scale events, or is it true on "all" scales? The dissipation is certainly intermittent on all scales that can be adequately resolved in space and time, and Occams principle thus speaks in favor of Parkers "nano-flares" relative to any mechanism that assumes nature hides another mechanism below the resolution limit.

The spatial and temporal complexity of the solar magnetic field is the focus of a rapidly growing literature. Magnetograms from plage regions show that the vertical component of the magnetic field has a self-similar horizontal distribution with a fractal dimension of around 1.5[1]. Both the spatial distribution and the temporal displacements of identifiable magnetic features are consistent with random walk of the magnetic field in the labyrinth of paths between convection cells on the solar surface[2]. The fractal distribution of the foot points of the coronal magnetic field makes it natural to assume that the coronal magnetic field itself has a fractal three-dimensional structure[3,4].

The spatial complexity is mirrored in a correspondingly complex temporal behavior. X-ray observation of flares by the Solar Maximum Mission show that the peak flare energy follows a power law distribution over at least 3.5 orders of magnitude in energy[5]. Radio observations show that the initial phase of flares is composed of many small scale millisecond events [6]. Based on numerical experiments, Lu and Hamilton[7] suggest that the flare energy power law distribution occurs because the magnetic field is in a self organized critical state[8,9]. This means that the field evolves on the edge of an unstable situation, where infinitesimal disturbances may trigger events of arbitrary size within the power law distribution, and that flare events are composed of a huge number of small events with physical sizes near the shortest length scale in the field.

The trend away from overly simplified scenarios ("text-book magnetic fields"; flux tubes, current sheets, simple arcades, etc.) is due to on the one hand necessity—observations force us to accept that reality is complex—on the other hand capability—numerical simulations have opened up the possibility to study reasonably complex systems semi-empirically. Computer speeds and memory capacities have reached levels where numerical experiments in some cases can be compared directly with observations and in other cases can provide important qualitative clues to understanding the observations.

In the subsequent Sections, we discuss three areas where numerical experiments have contributed to a qualitative understanding of some of the mechanisms mentioned above. In Section 2, we summarize current understanding of the topology of solar convection. In Section 3, we discuss numerical simulations of small scale magnetic fields at the solar surface, in Section 4 a recent numerical experiment that elucidates the dynamo process, and in Section 5 ideas and methods for modeling a spatially complex coronal magnetic field.

2. CONVECTION

Using numerical simulations with a realistic solar equation of state and a detailed treatment of radiative energy transfer, at a numerical resolution of $63 \times 63 \times 63$, Stein & Nordlund[10] discovered that the topology of solar convection below the solar surface is qualitatively different from the cellular appearance (granulation) at the visible surface. More recent results, obtained at higher spatial resolution[11,12] verify the earlier results, and are in very good agreement with high resolution observations of the solar surface. Fig. 1 shows a comparison of results from the numerical simulations with observations from the Swedish Solar Observatory at La Palma[13]. When smoothed by an appropriate point spread function, the intensity image from the numerical simulations (Fig. 1b) is remarkably similar to the observations (Fig. 1c). There is also agreement on the magnitude of the rms intensity fluctuation, which is about 8% in both the observed and the simulated (and smeared) image. The "true" intensity rms (Fig. 1a) is around 17% (bolometric), or about 20% at 5000 Å.

The subsurface behavior is illustrated in Fig. 2, which shows a 3-D rendering of surfaces of constant kinetic energy, relative to the maximum at each depth. In contrast to the cellular surface topology, the subsurface layers are dominated by thin filaments of cool gas, descending from the surface in a back-

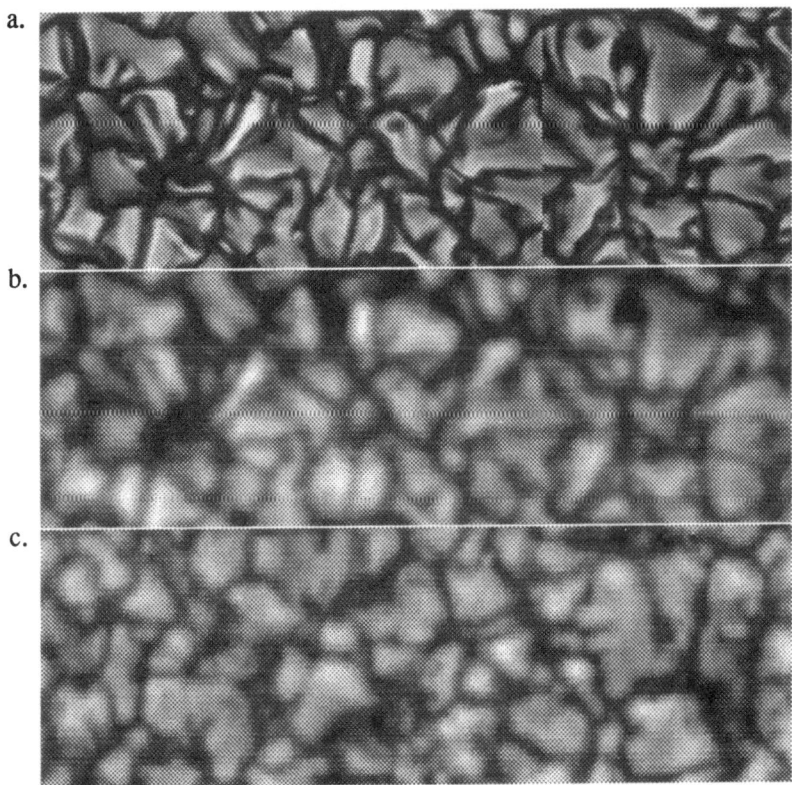

Fig. 1. Comparison of the intensity pattern from a numerical simulation of the solar granulation with observations from the Swedish Solar Observatory on La Palma. a) Three snap shots, separated by 5 minutes in time showing the surface intensity from a 6 × 6 × 3 Mm simulation (horizontal resolution 125 × 125). b) The same pattern, smoothed with a point spread function representative of the finite instrumental resolution and atmospheric seeing (applied to the composite image, thus hiding the sharp borders between adjacent frames). c) An area of the same size (6 × 18 Mm) from a 6302Å slit jaw image obtained at the Swedish Solar Observatory[13].

ground of slowly ascending, very nearly isentropic gas. In addition, the slow overturning on successively larger scales at larger depth advects the filaments horizontally and causes small scale (granular) filaments to merge into larger and more widely separated filaments on meso- and super-granular scales.

This particular topology arises because of the rapid increase of density with depth, and because the surface is the only source of significant cooling. The cool and thin filaments mark the location of gas that has been cooled at the surface, or that has mixed with cool gas from the surface.

16 Large Scale Simulations

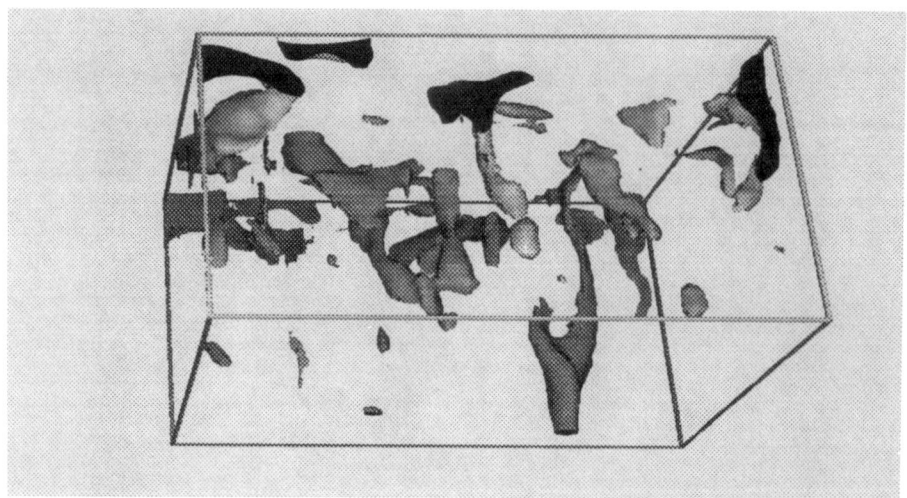

Fig. 2.   Perspective view (from front-left-below) of surfaces of constant relative kinetic energy, for a snapshot from a 6 × 6 × 3 Mm (resolution 63 × 63 × 63) simulation[10]. The structures show where the kinetic energy is larger than 50 % of its horizontal maximum value. The vertically oriented (lower) parts correspond to filaments with predominantly vertical velocities, while the horizontally oriented (upper) parts correspond to predominantly horizontal velocities near intergranular lanes at the surface.

3. SMALL SCALE SURFACE MAGNETIC FIELDS

Below the solar surface, the gas pressure is high enough to dominate the magnetic pressure. The solar surface and photosphere is a boundary layer, where the magnetic field gradually becomes dynamically significant. In the very surface layers, the surface cooling is so effective that the interiors of the thin vertical flux tubes that permeate the solar surface are cooled significantly below the average surface temperature. As a consequence, the pressure scale height is smaller than in the surrounding photosphere, and the internal pressure at vertical pressure equilibrium is therefore significantly smaller than the external pressure. Horizontal pressure equilibrium obtains by the additional magnetic pressure of the flux tube which adds to the internal gas pressure. Current simulations[14,15] do not yet have sufficiently high resolution to show the full amount of evacuation of photospheric magnetic flux tubes. They do, however, illustrate the significant influence on the shape and size of the granulation pattern, which has smaller and more roundish granules in the presence of a magnetic field, in accordance with observations[16]. The modified granulation pattern results in a modification of the energy transfer through the solar surface, with a corresponding change of the mean temperature structure. For not too large filling factors, the temperature (at constant pressure) actually *increases* in the thin surface layer[12].

The intermittency of magnetic fields at the solar surface implies that the field is rather loosely coupled to the solar plasma as long as the horizontal filling factor is small. For large enough filling factors, the surface magnetic field forms a connected pattern and the magnetic field is much more tightly coupled to the plasma[17].

1. DYNAMO ACTION

The ultimate source of the magnetic fields that penetrate the solar surface is the solar dynamo process, which generates the quasi-cyclic global solar magnetic field. The solar dynamo is widely believed to be driven by the interaction of large scale convection and the solar (differential) rotation, with the primary site of the dynamo action located near the interface between the convection zone and the stable layers below the convection zone[18-20] The qualitative effects involved in this interaction were explored in a numerical experiment[21] designed to model a limited region near the lower boundary of a convection zone similar to the solar one. By choosing a relatively small Cartesian volume one sacrifices the global aspects of the solar dynamo process, in order to make it possible to resolve the strongly anisotropic (filamentary) nature of the convective motions. Rotation is accounted for by including a Coriolis force with a constant rotation vector Ω. In this particular experiment, the angle between Ω and the acceleration of gravity g was 60°, corresponding to a latitude of 30°.

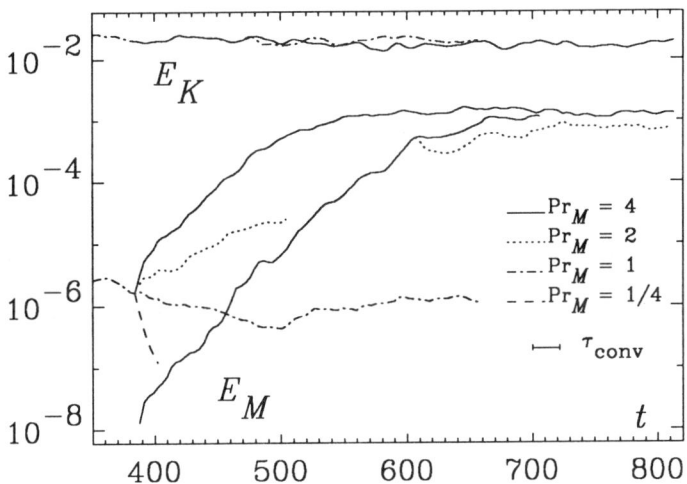

Fig. 3. The exponential growth of magnetic energy E_M and subsequent saturation (lower two solid curves, $Pr_M = 4$). E_K is the kinetic energy. A reduction by a factor 10^2 in the seed fields strength gives the same saturation energy (lower solid curve, $Pr_M = 4$). For smaller Pr_M there is no growth (lower dotted-dashed curve, $Pr_M = 1$).

This simple setup is sufficient for spontaneous dynamo action. A week seed field with vanishing average value (introduced after convection has been allowed to relax) grows exponentially with time, until saturation occurs (Fig.

18 Large Scale Simulations

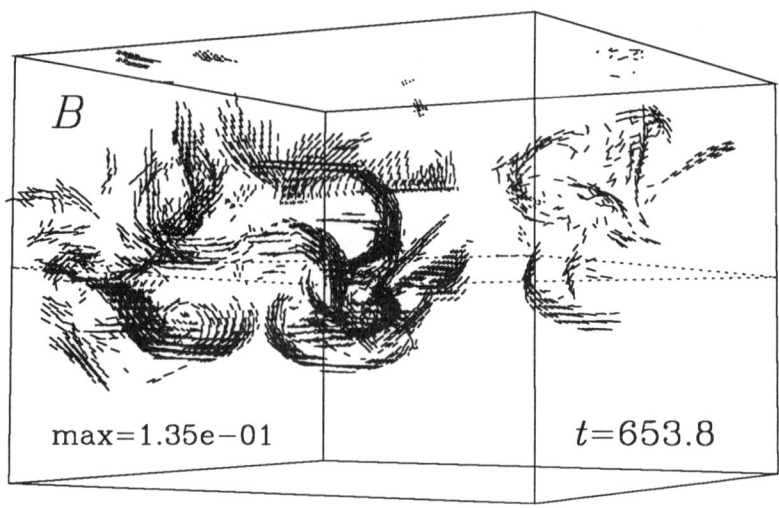

Fig. 4. Magnetic field vectors in a 3-D perspective plot for a snapshot from the dynamo simulation[21]. Only vectors above a certain threshold are plotted. The interface between stably and unstably stratified layers is marked by a dotted line.

3). Most of the magnetic energy is concentrated in thin magnetic flux tubes with diameters corresponding to only a few mesh points (Fig. 4). The saturated value of the magnetic energy is independent of the amplitude of the seed field (cf. the lowermost full drawn curve in Fig. 3). The efficiency of the dynamo action depends sensitively on the magnetic Prandtl number $Pr_M = \nu/\eta$; the ratio of the kinematic viscosity and the magnetic diffusivity. The magnetic Prandtl number measures the importance of viscous drag on a flux tube relative to diffusion of the flux tube through the plasma. Dynamo amplification occurs because magnetic flux tubes are repeatedly sucked into the downdrafts, twisted by the helical motion (which in turn is due to rotation), stretched by the expansion of descending fluid as it hits the boundary to the stable layers, and folded back into the downdraft by overturning ascending fluid. Saturation occurs when the Lorentz force is strong enough to force the flux tubes to slip through the plasma rather than stretching and bending further. Splitting the Lorentz force into pressure, tension and curvature forces, one finds that at saturation it is the work done against the curvature force that balances the resistive dissipation[21].

The process is qualitatively similar to the idealized stretch-twist-fold scenario of Vainshtein & Zeldovich[22]; it is the possibility of three-dimensional foldings that enables dynamo action. Note that in non-trivial three-dimensional models (as in reality) magnetic field lines do not close upon themselves.

5. SELF-SIMILAR MAGNETIC FIELDS

As discussed in the Introduction, there is strong evidence for a self-similar (fractal) three-dimensional distribution of the coronal magnetic field. What would characterize the topology of such a field? In fluid mechanics, interest has recently focussed on the importance of critical points (stagnation points)[23,24]. The corresponding concept for magnetic fields is magnetic null points[25,26], which are points where all three components of the magnetic field vanish. For a general 3-D magnetic field, with no particular symmetries, one component of the field vanishes on a set of surfaces, two components vanish on a set of lines, and all three components vanish on a set of points. Sufficiently complex magnetic field may thus be expected to contain null-points. If the magnetic field is fractal, the distribution of null-points is fractal as well. Separator surfaces associated with each null-point determine the local connectivity of magnetic field lines and are topologically robust; only the pairwise destruction or creation of null points changes the associated connectivity[26].

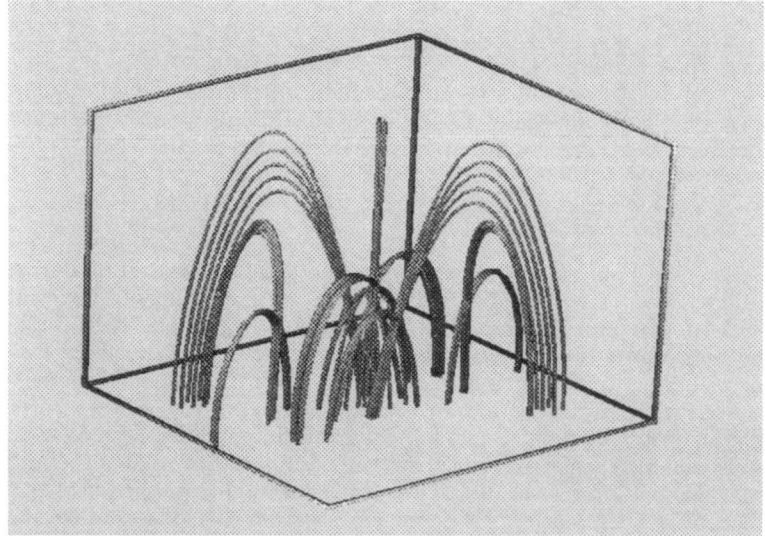

Fig. 5. A 3-D rendering of a magnetic field with two wavevector sets, derived from a scalar potential with exponential z-dependence and sinusoidal xy-dependence. The wavenumbers of the smaller scale set are four times those of the larger scale set, and the amplitude ratio at $z = 0$ is 1:6.

Magnetic fields with a rudimentary self-similarity may be constructed by using a "sparse" Fourier series, with a power law distribution of Fourier amplitudes[3,4,27]. Most random combinations of Fourier components with similar but distinct wavelengths result in magnetic fields with null points separated by scales comparable to the wavelength of the Fourier components. Additional Fourier components with shorter wavelengths lead to the creation of "satellite" null points in the neighborhood of the larger scale null points. Continuing the process recursively one obtains a magnetic field with a selfsimilar set of null points.

Fig. 5 shows a 3-D rendering of a potential magnetic field consisting of two sets of similar Fourier components. The box is centered on a null point of

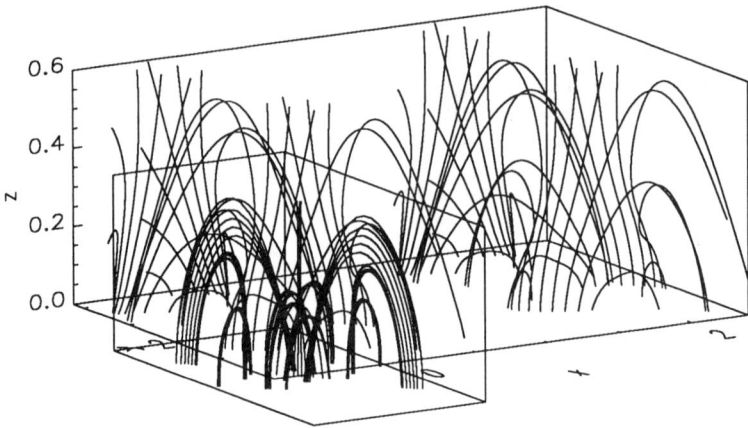

Fig. 6. 3-D perspective plots of field line traces. The larger box corresponds to the scale of the larger scale Fourier components and the inset box corresponds to the scale of the smaller scale Fourier components.

the largest scale Fourier component set. The smaller scale Fourier component set is able to create five "satellite" null-points in the vicinity of the central null point, but has virtually no influence on the topology of the field further away from the null points (because of the smaller amplitude of the short wavelength variation). This is further illustrated in Fig. 6, which shows field line traces in a larger box, with an inset box corresponding roughly to the box in Fig. 5.

The extent to which a smaller wavelength set is able to create satellite null-points depends sensitively on the spectral index that relates the amplitudes of fluctuations on different scales. The more rapidly the amplitude of the smaller scale fluctuations decrease with decreasing wavelength, the smaller is the volume around the larger scale null point where the smaller scale fluctuations are felt. This is illustrated in Fig. 7, which shows a surface plot of the inverse of the magnetic field strength (mapped to the unit interval) in an xy plane, for three different spectral indices. The figure illustrates the strong influence of the spectral index on the field topology.

Field line traces of the two extreme cases are shown in Fig. 8. Only two wavelength sets are used in this plot. The three wavelength set field (cf. Fig. 7) is difficult to visualize on this scale, since it has more than 40 null points in the box for the case with spectral index = -1.58.

6. CONCLUDING REMARKS

The sensitivity of the topology of a self-similar magnetic field to the amplitude spectral index suggests a possible scenario for the self-organized critical state suggested by Lu and Hamilton[7]. Boundary stress from the photospheric foot point motion tends to increase the complexity of the coronal magnetic field, which corresponds to a lowering of the amplitude spectral index and an increase in the number of critical points in the field. Stressed but force-free three-dimensional magnetic null points are sources of local instabilities, for much the

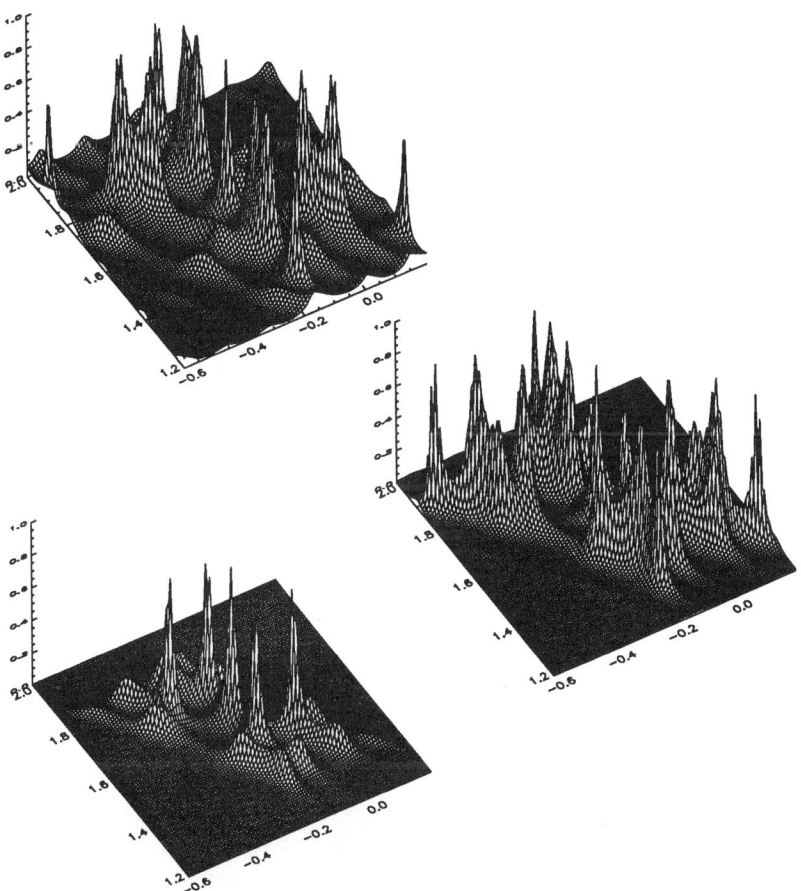

Fig. 7. Surface plot of three magnetic fields with different power law distributions of the amplitude vector. Each field consists of three wavelength sets, with relative sizes 1, 1/4 and 1/16. The three figures have scalar potential spectral indices equal to -1.58, -1.66 and -1.78, respectively.

same reason that magnetic X-points can drive instabilities in two dimensions[28]. When the stress exceeds the limit for local stability, small scale instabilities are triggered that release some of the stress and dissipate magnetic energy.

Such a system would achieve quasi-static energy balance by developing a state of self-organized criticality, where the spatial complexity always remains near a level where local instabilities are easily triggered. Such a state is characterized by a power law distribution of event sizes, with relatively rare but powerful large scale events and a large number of smaller scale events, all the way down to the smallest (resistive) scales. If the individual events deposit some of the dissipated energy in the form of accelerated particles, the scenario fits in

22 Large Scale Simulations

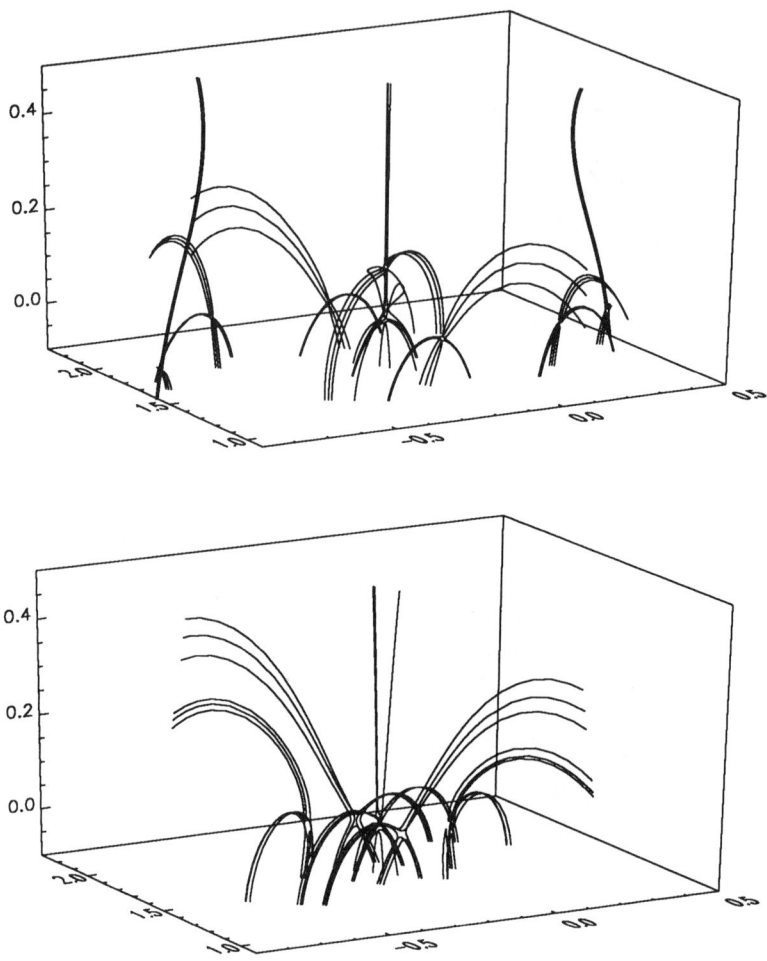

Fig. 8. Field line traces of fields with power law distribution spectral indices of -1.58 (upper) and -1.78 (lower). The fields consist of two similar wavelength sets, with a size ratio of 1:4.

nicely with recent theories of flare particle acceleration[29].

This work was supported by grants from the Danish Natural Science Research Council, the Danish Space Board, and the Carlsberg Foundation.

REFERENCES

1. T. Tarbell, D.S. Acton, K. Topka, A. Title, W. Schmidt, G. Scharmer, Phys. Fluids , submitted (1991).

2. C.J. Schrijver, C. Zwaan, A.C. Balke, T.D. Tarbell, J.K. Lawrence, Letter to Astronomy Astrophys , (in press) (1991).
3. K. Galsgaard, Å. Nordlund, Proc. Heidelberg Conference on: Mechanisms of Chromospheric and Coronal Heating, Eds. P. Ulmschneider, E.R. Priest, R. Rosner, (Springer, 1991) p. 541.
4. K. Galsgaard, Å. Nordlund, IAU Coll. 130, The Sun and Cool Stars: Activity, Magnetism, Dynamos, Eds. I. Tuominen, D. Moss, G. Rüdiger, Lecture Notes in Physics, (Springer, 1991) p. 89.
5. B.R. Dennis, Solar Phys. 100, 465 (1985).
6. A.O Benz, M.J. Aschwanden, IAU Coll. 133, Eruptive Solar Flares, Eds. B.V. Jackson, M.E. Machado, Z.F. Svestka, (Springer, 1992).
7. E.T. Lu, R.J. Hamilton, Astrophys. J. Letters 380, L89 (1991).
8. P. Bak, K. Chen, K. Wiesenfeld, Physical Review A 38, 364 (1988).
9. P. Bak, K. Chen, Scientific American Jan, 26 (1991).
10. R.F. Stein, Å. Nordlund, Astrophys. J. Letters 342, L95 (1989).
11. Å. Nordlund, R.F. Stein, Challenges to Theories of the Structure of Moderate Mass Stars, Eds D.O. Gough, J. Toomre, Lecture Notes in Physics, Vol. 388, (Springer, 1991) p. 141.
12. Å. Nordlund, R.F. Stein, NATO Advanced Research Workshop on: Stellar Atmospheres: Beyond Classical Models, NATO ASI Series C, Vol 341., (Kluwer, 1991) p. 263.
13. B.W. Lites, Å. Nordlund, G.S. Scharmer, Solar and Stellar Granulation, Eds R.J. Rutten, G. Severino, (Kluwer, 1989) p. 349.
14. Å. Nordlund, R.F. Stein, Solar and Stellar Granulation, Eds R.J. Rutten, G. Severino, (Kluwer, 1989) p. 453.
15. Stein, R.F., A. Brandenburg, Å. Nordlund, Proceedings of the Seventh Cambridge Workshop on Cool Stars, Stellar Systems and the Sun, Eds. M. S. Giampapa and J. A. Bookbinder, (ASP Conf. Series, 1991).
16. A. Title, Proceedings of the Seventh Cambridge Workshop on Cool Stars, Stellar Systems and the Sun, Eds. M. S. Giampapa and J. A. Bookbinder, (ASP Conf. Series, 1991).
17. Å. Nordlund, R.F. Stein, IAU Symp. 138, Solar Photospher: Structure, Convection and Magnetic Fields, Ed. J.-O. Stenflo, (Kluwer, 1990) p. 189.
18. E.N. Parker, Astrophys. J. 198, 205 (1975).
19. E.A. Spiegel, N.O. Weiss, Nature 287, 616 (1980).
20. H.C. Spruit, A.A. van Ballegooijen, Astron. Astrophys. 106, 58 (1982).
21. Å. Nordlund, A. Brandenburg, R.L. Jennings, M. Rieutord, J. Roukolainen, R.F. Stein, I. Tuominen, Astrophys. J. , submitted (1992).
22. S.I. Vainshtein, Ya.B. Zeldovich, Sov. Phys. Usp. 15, 159 (1972).
23. A.E. Perry & M.S. Chong, Ann. Rev. Fluid. Mech 19, 125 (1987).
24. M.S. Chong, A.E. Perry, B.J. Cantwell, Proceedings of the IUTAM Symposium on Topological Fluid Mechanics, Eds. H.K. Moffatt, A. Tsinober, (Cambridge University Press, 1990) p. 408.
25. S. Fukao, U. Masayuki, T. Takao, Rep. Ionos. Res. Jpn. 29, 133 (1975).
26. J.M. Green, J. Geophys. Res. 93, 8583 (1989).
27. K. Galsgaard, Å. Nordlund, IAU Coll. 133, Eruptive Solar Flares, Eds. B.V. Jackson, M.E. Machado, Z.F. Svestka, (Springer, 1992).
28. K. Galsgaard, Topology of Magnetic Fields in Three Dimensions, masters thesis, (Copenhagen University, 1991).
29. L. Vlahos, Solar Phys. 121, 431 (1989).

MAGNETIC FLUX TUBES AS COMMUNICATION CHANNELS

B. Roberts
Department of Mathematical and Computational Sciences,
University of St Andrews, St Andrews, Fife KY16 9SS, Scotland

ABSTRACT

The role of magnetic flux tubes as communication channels connecting the photosphere with the chromosphere and above is reviewed.

INTRODUCTION

The photosphere and upper convection zone is a dynamic place. It is here that motions are observed, with amplitudes of 1-3 km s^{-1} in granules and about 1/2 km s^{-1} in supergranules, and it is here that the occasionally more dramatic motion occurs in 'exploding' granules. It is here too that the roots of the chromospheric and coronal magetic field are to be found. A hierarchy of magnetic flux concentrations occur, supporting the magnetic field that higher in the atmosphere expands out to fill the available space. The hierarchy of magnetism ranges from the small scale intense magnetic flux tubes to the large scale sunspots, with pores and knots providing intermediate structures. The small scale isolated magnetic flux tubes, located in the downdraughts between supergranules and bordered by granules, are subject to the dynamic motions that surround them. In consequence, we may anticipate that these tubes will be jossled about and generally manipulated by the flows and wave fields in their environment.

Magnetic flux tubes are communication channels: they link the high energy density part of the solar atmosphere, the photosphere, with the tenuous low energy density but hot higher reaches of the atmosphere in the chromosphere and corona. This linkage is essentially one-dimensional as flows and waves are guided by the magnetic field, though the efficiency of this guide diminishes with height as the field expands out. It is natural, then, to enquire into the detailed nature of magnetic flux tubes and the waves they support. We present an overview of this topic here.

What is a photospheric magnetic flux tube? In its simplest form, we may view it as a cylindrical concentration of magnetic field-lines of strength B_0, though slab configurations are also of interest for modelling accumulations of magnetic flux in intergranular lanes. Total pressure balance implies that the tube is a region of low gas pressure and, unless the thermal structure is highly unusual, it is also a region of gas depletion:

$$p_0 + \frac{B_0^2}{2\mu} = p_e, \qquad \rho_0 < \rho_e. \tag{1}$$

Here p_0 and ρ_0 are the gas pressure and density within the tube which is embedded within a field-free environment of pressure p_e and density ρ_e; μ is the magnetic permeability of the gas. The partial evacuation of gas within a tube is controlled in part by its temperature structure. Detailed models of photospheric flux tubes reveal a complicated thermal structure, with the lower subphotospheric layers being cooler than their surroundings while the upper layers are warmer than the environment (see the review by Schussler[1]); nonetheless, tubes remain regions of density depletion.

Gravitational stratification causes a tube to expand outwards. Consequently, at some level in the chromosphere neighbouring tubes merge and a canopy structure is formed, providing a magnetic cover for the upper convection zone. Within the expanded upper reaches of the tubes spicules occur–jets of gas that are observed to rise with speeds of about 25 km s^{-1} into the corona. The canopy, separating the field–dominated upper atmosphere of the sun from the almost field-free lower layers, is evidently a complicated site where flows and sound waves must generate and interact with magnetoacoustic disturbances.

Both the canopy and the concentrated magnetic flux tubes that support it are, of course, not rigid structures but are elastic and so able to support waves of various kinds. Shock waves, too, are likely to impinge on these structures; shock waves may be associated with 'exploding' granules. Certainly computational simulations reveal shock structures associated with granular motions.[2] However, no detailed theoretical studies of the interactions of shocks with either canopy structures or magnetic flux tubes have been carried out, though studies of shocks within tubes have been made.[3–5] As well as the possibility of shocks occurring in magnetic flux tubes, theoretical studies have shown that solitons may arise.[6–11] Solitons are nonlinear waves that propagate almost as particles, preserving their identities after interactions. They owe their existence to the combined influences of nonlinearity and dispersion; the nonlinearity is the usual one of any gas, the dispersion arising because wave motions within the tube tend to disturb, albeit only slightly, the environment of the tube and the extent of this disturbance depends upon the wavelength of the wave.

The ability of a tube to support solitons is associated, then, with the tube having a length-scale, separate from any that the disturbance itself may possess. (This is in contrast with the case of a uniform medium, which has no length-scale save that determined by the perturbation itself; consequently, magnetohydrodynamic waves in a uniform medium are non-dispersive.)

As well as guiding wave motions, tubes may also leak waves into their environment and, furthermore, absorb them from the environment. The leakage of waves is associated with the generation of sound in the environment of the tube, with the sound wave propagating laterally away from the tube to infinity. The reverse process, by which sound waves are absorbed within the tube, is presumably the basis for the observed strong absorption of p-modes by sunspots,[12] though a detailed theoretical explanation has yet to be accepted.

THE BASIC MODES OF OSCILLATION OF A TUBE

Consider a gas of constant density ρ_0 and pressure p_0 within which is embedded a uniform magnetic field $\boldsymbol{B}_0$ of strength B_0. The two fundamental speeds of this system are the Alfvén and sound speeds, c_A and c_s, defined by

$$c_A = \left(\frac{B_0^2}{\mu\rho_0}\right)^{1/2}, \qquad c_s = \left(\frac{\gamma p_0}{\rho_0}\right)^{1/2}, \qquad (2)$$

for adiabatic index γ.

The Alfvén speed is the characteristic speed with which torsional oscillations of the tube are propagated. Any squeezing of the tube, however, can give rise to sausage and kink modes. The sausage mode arises whenever the tube is disturbed in a symmetric fashion, with the main axis of the tube remaining unperturbed. The sausage mode propagates with the speed c_t, defined by

$$\frac{1}{c_t^2} = \frac{1}{c_s^2} + \frac{1}{c_A^2}; \qquad (3)$$

evidently c_t is both sub-sonic and sub-Alfvénic. If the sound and Alfvén speeds are widely separated, then the speed c_t closely approximates the smaller of the two speeds.

A speed equivalent to c_t is in fact common to a variety of elastic tubes, with the role of the Alfvén speed being played by the appropriate elastic speed of the physical situation. For example, in the case of a blood vessel the speed c_A is simply the elastic speed in the membrane of the blood vessel. In this case, the speed of sound in blood is so large that $c_s \gg c_A$ and so the sausage mode in a blood vessel propagates with the speed $c_t \approx c_A$, the elastic speed in the membrane of the vessel. In the case of 'water hammer' in a pipe, the relative magnitudes of the two basic speeds can interchange with a change in the material of the pipe.[13] In a metal pipe, the elastic speed c_A of the metal membrane is much larger than c_s, the speed of sound in water, and so the speed $c_t \approx c_s$ (which, incidentally, is about 1.4 km s^{-1}, some 20% of the sound speed in the photosphere). In a plastic pipe, the inequality is reversed and $c_A \ll c_s$, and so $c_t \approx c_A$ (which is about 10 m s^{-1} for a plastic pipe).

Finally, we consider asymmetric distortions of the tube which produce kink (or serpentine) modes. These can be thought of as similar to the vibrations of an elastic string satisfying the wave equation

$$\rho \frac{\partial^2 \xi}{\partial t^2} = T \frac{\partial^2 \xi}{\partial z^2}, \qquad (4)$$

for lateral displacement ξ, tension T and density ρ. The wave speed is accordingly $(T/\rho)^{\frac{1}{2}}$. We may associate the tension T with the magnetic tension force B_0^2/μ. The density ρ, however, is comprised of two contributions, $\rho_0 + \rho_e$, because the tube displaces about the same amount of fluid in its surroundings as within the tube itself. Consequently, kink disturbances are characterized by the speed[14-16,26]

$$c_k = \left(\frac{\rho_0}{\rho_0 + \rho_e}\right)^{\frac{1}{2}} c_A, \qquad (5)$$

which is sub-Alfvénic.

To understand the oscillations of a tube in greater detail it is necessary to examine the Fourier form of its modes of oscillation. Consider, then, an isolated flux tube of radius a, the axis of which is aligned with the z-axis of a cylindrical coordinate system (r, θ, z). The equilibrium magnetic field and gas density are taken to be of the form

$$B_0(r) = \begin{cases} B_0, & r < a, \\ 0, & r > a, \end{cases} \qquad \rho_0(r) = \begin{cases} \rho_0, & r < a, \\ \rho_e, & r > a. \end{cases} \qquad (6)$$

The undisturbed tube is in pressure balance, as described by equation (1). Combined with the ideal gas law, pressure balance implies that the densities within and external to the tube are related by

$$\frac{\rho_e}{\rho_0} = \frac{c_s^2 + \frac{1}{2}\gamma c_A^2}{c_e^2}, \qquad (7)$$

where c_e denotes the speed of sound in the environment ($r > a$) of the tube.

Normal modes of oscillation are described by writing

$$f(r, \theta, z, t) = f(r) \exp i(\omega t + n\theta - k_z z), \qquad (8)$$

for each perturbation quantity $f(r, \theta, z, t)$. The geometry of the modes is then described by the integer n: $n = 0$ describes the sausage mode; $n = 1$ describes the kink mode; and $n \geq 2$ gives the fluting modes. It is necessary also to distinguish between surface modes and body modes. Surface modes are non-oscillatory in r both outside and within the tube, whereas body modes are non-oscillatory outside the tube but are oscillatory in r within the tube.[17]

With the prescription (8) it is straightforward to analyze the behaviour of linear disturbances, obtaining as a result the dispersion relation for magnetoacoustic waves in an isolated magnetic flux tube (see Edwin and Roberts[18] and references therein for details):

$$\rho_e \omega^2 n_0 \frac{J'_n(n_0 a)}{J_n(n_0 a)} + \rho_0 (k_z^2 c_A^2 - \omega^2) m_e \frac{K'_n(m_e a)}{K_n(m_e a)} = 0, \qquad (9)$$

where

$$n_0^2 = \frac{(\omega^2 - k_z^2 c_A^2)(\omega^2 - k_z^2 c_s^2)}{(c_s^2 + c_A^2)(\omega^2 - k_z^2 c_t^2)}, \qquad (10)$$

and

$$m_e^2 = k_z^2 - \frac{\omega^2}{c_e^2}. \qquad (11)$$

In the above, J_n and K_n are the Bessel and modified Bessel functions of order n, and a dash (') denotes the derivative of the Bessel function (e.g., $J'_n(n_0 a) \equiv dJ_n(x)/dx$ evaluated at $x = n_0 a$, etc.).

In the derivation of the dispersion relation (9) it is assumed that m_e^2 is positive, corresponding to selecting waves that decay (in r) away from the tube. By contrast, n_0^2 may be positive or negative; waves with n_0^2 positive are the body modes, whereas waves with n_0^2 negative are the surface modes.

It is interesting to compare equation (9) with the usual dispersion relation for magnetoacoustic waves in an unbounded (unstructured) medium:

$$\omega^4 - k^2(c_s^2 + c_A^2)\omega^2 + k^2 k_z^2 c_s^2 c_A^2 = 0, \qquad (12)$$

where $k^2 = k_z^2 + k_\perp^2$ for perpendicular wavenumber $k_\perp$. In fact, the dispersion relation (12) of a uniform medium may be rewritten in the form

$$k_\perp^2 = n_0^2, \tag{13}$$

from which it is evident that only body modes (the familiar fast and slow magnetoacoustic waves) occur; see Roberts[19] for a recent discussion of magnetohydrodynamic waves in a uniform medium.

The two types of solution of the dispersion relation (9), corresponding to the body and surface waves, require a detailed, largely numerical, investigation to extract them from its transcendental structure.[18,20] The topic has been reviewed recently in Roberts[21] and so we will be content here merely to recap on the basic findings. There are body modes ($n_0^2 > 0$) with phase-speeds ω/k_z that lie between c_t and c_s and correspond to slow modes. The slow body modes may be viewed as waves that are constrained within the tube, bouncing from side to side of the tube as they propagate along its interior. There are also slow surface waves which have phase-speeds along the tube that are less than c_t.

Finally, there are fast modes the nature of which depends upon the relative orderings of the various basic speeds. For example, when $c_e > c_s > c_A$ the fast modes are body waves with their various harmonics; this ordering of speeds is appropriate for the sub-photospheric layers of a sunspot, for example. By contrast, when $c_A > c_e > c_s$ the fast waves are surface modes; this ordering of speeds is appropriate for the upper layers of a sunspot. The change in character in the modes when the basic speeds are differently ordered suggests that sunspots, for example, may support p-modes (fast waves) in the lower layers but in the upper layers these modes are unable to propagate but will leak from the spot or couple into surface modes (see Evans and Roberts[20]). Slow body modes, however, which exist in either ordering of the speeds, will propagate from the lower layers of a spot to its upper atmosphere.

So far in our discussion we have treated the case of no gravity. However, in the photosphere-chromosphere of the sun gravity has important consequences that need to be assessed. For wide tubes, such as sunspots, this has not yet proved entirely possible, though for thin tubes progress has been possible within the thin tube approximation to which we now turn.

THE THIN TUBE EQUATIONS

The first effect of gravity on an isolated flux tube is to render it non-uniform in height. The fall-off of the confining gas pressure in the environment of the tube forces the tube to expand outwards. The effect is most easily demonstrated for thin tubes, that is for tubes that vary little across their structure.

Considering tubes that are symmetric about the z-axis of a cylindrical coordinate system, we may describe thin tubes by the so-called thin tube equations for the sausage mode:

$$\frac{\partial}{\partial t}\rho A + \frac{\partial}{\partial z}\rho v A = 0, \tag{14}$$

$$\frac{\partial v}{\partial t} + v\frac{\partial v}{\partial z} = -\frac{1}{\rho}\frac{\partial p}{\partial z} - g, \tag{15}$$

$$\frac{\partial p}{\partial t} + v\frac{\partial p}{\partial z} = \frac{\gamma p}{\rho}\left(\frac{\partial \rho}{\partial t} + v\frac{\partial \rho}{\partial z}\right), \qquad (16)$$

$$p + \frac{B^2}{2\mu} = \pi_e, \qquad (17)$$

$$BA = \text{constant}. \qquad (18)$$

In these equations, $B(z,t)$ is the field-strength of the tube, $A(z,t)$ its cross-sectional area, $v(z,t)$ the longitudinal flow speed within the tube, where the gas pressure and density are $p(z,t)$ and $\rho(z,t)$. The external gas pressure $\pi_e(z,t)$ is calculated on the boundary of the tube. We have chosen the z-axis to be aligned with gravity, pointing upwards. The thin tube equations have been derived in Roberts and Webb[22] by expanding in Taylor series the equations of ideal magnetohydrodynamics about the central axis ($r = 0$) of the tube (see also Refs. 23-25). Thin tube equations have also been derived for the kink mode; see Spruit.[26]

Notice that the five equations which comprise the thin tube equations for the sausage mode (in fact, slow surface sausage mode) of a tube may be grouped into two separate halves: equations (14)-(16) refer to any thin elastic tube, whereas equations (17) and (18) describe the magnetic elasticity of a magnetic flux tube. For other tubes, such as the hose-pipe mentioned in the Introduction (see also Lighthill[13] and Campos[27]), a different set of equations replaces (17) and (18) to describe the way in which the cross-sectional area of the tube responds to pressure variations.

In the equilibrium ($v = 0$) configuration, the thin tube equations yield

$$p_0(z) = p_0(0)e^{-N}, \qquad \rho_0(z) = \rho_0(0)\frac{\Lambda_0(0)}{\Lambda_0(z)}e^{-N}, \qquad (19)$$
$$A_0(z) = A_0(0)e^{N/2}, \qquad B_0(z) = B_0(0)e^{-N/2},$$

where

$$N(z) = \int_0^z \frac{dz}{\Lambda_0(z)}, \qquad \Lambda_0(z) = \frac{p_0(z)}{g\rho_0(z)} \qquad (20)$$

and it has been assumed that the pressure scale-height $\Lambda_0(z)$ inside the tube is equal to that outside the tube. The environment is taken to be in hydrostatic equilibrium.

When gravity is negligible ($g = 0$), it proves possible to extract from the thin tube equations the behaviour of weakly nonlinear, weakly dispersive disturbances. The slow sausage wave of a thin tube (for which $\omega \approx k_z c_t$) is governed by an integro-differential equation of the form (see Roberts[7]; Molotovshchikov and Ruderman[8]).

$$\frac{\partial v}{\partial t} + c_t\frac{\partial v}{\partial z} + \beta_0 v\frac{\partial v}{\partial z} + \alpha_0 \frac{\partial^3}{\partial z^3}\int_{-\infty}^{\infty}\frac{v(s,t)ds}{[\lambda^2 a^2 + (z-s)^2]^{1/2}} = 0, \qquad (21)$$

where $\lambda = c_t/c_A$ (for $c_s = c_e$); the constants α_0 and β_0 are complicated coefficients not given here (see Roberts[7]). In the above, $v = v(z,t)$ denotes the flow velocity along the central axis of the tube. Equation (21), sometimes referred to as the Leibovich-Roberts equation,[10] has been solved numerically[11] and exhibits solutions that appear to be soliton-like in character. Unfortunately, no analytical solution of equation (21) is presently known. However, the equivalent problem for a slab of magnetic field, rather than a cylindrical tube, is known to lead to solitons; the governing equation, replacing (21), is the Benjamin-Ono equation[6,7,28], namely

$$\frac{\partial v}{\partial t} + c_t \frac{\partial v}{\partial z} + \beta_0 v \frac{\partial v}{\partial z} + \alpha_1 \frac{\partial^2}{\partial z^2} \int_{-\infty}^{\infty} \frac{v(s,t)}{s-z} ds = 0, \quad (22)$$

for constant α_1. The Benjamin-Ono equation, involving the Hilbert transform (with Cauchy principal value integral), has been extensively studied.[29,30]

It is natural to wonder if spicules are manifestations of solitons in magnetic flux tubes.[6] Certainly there are some attractive similarities between the two phenomena, but the absence of gravity in the above theoretical description makes it difficult to develop a convincing case for spicules as solitons. One wonders how the diverging geometry of a flux tube as it expands into the chromosphere influences the propagation of a soliton, and whether this is consistent with the observational properties of spicules.

We turn now to a consideration of linear disturbances, for which it is possible to retain the effects of stratification ($g \neq 0$). It is usual to assume that the temperature inside the tube is equal to that in the environment, and also that the external pressure field $\pi_e(z,t)$ is equal to the undisturbed hydrostatic pressure $p_e(z)$ of the environment. Then, as shown by Rae and Roberts,[31] for any elastic tube (be it rigid or non-magnetic or magnetic) described by equations (14)-(16), linear disturbances are described by

$$\frac{\partial^2 Q}{\partial t^2} - c^2(z)\frac{\partial^2 Q}{\partial z^2} + \Omega^2(z)Q = 0, \quad (23)$$

where the amplitude $Q(z,t)$ is related to the flow $v(z,t)$ by

$$Q(z,t) = \left[\frac{\rho_0(z)A_0(z)c^2(z)}{\rho_0(0)A_0(0)c^2(0)}\right]^{1/2} v(z,t). \quad (24)$$

The speed c is determined by

$$\frac{1}{c^2} = \frac{1}{c_s^2} + \frac{\rho_0}{A_0}\left(\frac{\partial A}{\partial p}\right)_{p=0}, \quad (25)$$

and Ω^2 is given by

$$\frac{\Omega^2}{c^2} = \frac{1}{2}\left(\frac{\rho_0'}{\rho_0} + \frac{A_0'}{A_0} + \frac{c^{2\prime}}{c^2}\right)' + \frac{1}{4}\left(\frac{\rho_0'}{\rho_0} + \frac{A_0'}{A_0} + \frac{c^{2\prime}}{c^2}\right)^2 + \left(\frac{g}{c_s^2} - \frac{A_0'}{A_0}\right)' \\ + \left(\frac{g}{c_s^2} - \frac{A_0'}{A_0}\right)\left(\frac{\rho_0'}{\rho_0} + \frac{c^{2\prime}}{c^2} + \frac{g}{c_s^2}\right) - \frac{g}{c^2}\left(\frac{\rho_0'}{\rho_0} + \frac{g}{c_s^2}\right). \quad (26)$$

Thus the sausage mode in a thin tube stratified under gravity satisfies the Klein-Gordon equation.[31]

We may illustrate equation (26) for several special cases. For example, consider an isothermal atmosphere within a rigid tube with cross-sectional area $A_0(z) = A_0(0)\exp(\alpha z/\Lambda_0)$, similar to the area profile of a magnetic tube in an isothermal atmosphere. Then equation (25) gives $c = c_s$, and so (as expected) the wave propagates with the sound speed within the rigid tube. Equation (26) reduces to

$$\Omega^2 = \Omega^2_{rigid} = \left[\alpha^2 - (\frac{4}{\gamma} - 2)\alpha + 1\right]\omega_a^2, \quad (27)$$

where $\omega_a = c_s/(2\Lambda_0)$ is the acoustic cutoff frequency for an isothermal atmosphere. If the tube is straight, corresponding to $\alpha = 0$, then $\Omega_{rigid} = \omega_a$. If, however, the tube e-folds in 2 scale-heights, corresponding to $\alpha = 1/2$, then $\Omega_{rigid} = (\frac{9}{4} - \frac{2}{\gamma})^{\frac{1}{2}}\omega_a$, and the diverging rigid tube has a larger cutoff frequency (for $\gamma = 5/3$) than a straight tube.

Consider now a magnetic tube in an isothermal atmosphere. The equilibrium configuration (19) gives $A_0(z) = A_0(0)\exp(z/2\Lambda_0)$, corresponding to an effective value of $\alpha = 1/2$. Then equations (17) and (18), combined with equation (25) for c^2, show that $c = c_t$, and so the wave propagates with the slow magnetoacoustic speed c_t (as expected from the earlier discussion of the $g = 0$ case). The frequency Ω arising in equation (23) is given by[24,22]

$$\Omega^2 = \Omega^2_{sausage} = \left[\left(\frac{9}{4} - \frac{2}{\gamma}\right) - \left(\frac{3}{2} - \frac{2}{\gamma}\right)\frac{\beta}{\beta + (2/\gamma)}\right]\omega_a^2, \quad (28)$$

where $\beta = 2c_s^2/\gamma c_A^2$ is the plasma beta within the tube. Notice that this expression is made up of a term (the first on the right-handside of equation (28)) arising from the shape of the tube, as is clear by comparison with the case of a rigid tube, and a term (the second on the right-handside of (28)) which relates to the elasticity of the tube. In the case of a strong field, corresponding to low β, expression (28) reduces to the result for a rigid tube.

The Klein-Gordon equation arises also in describing the kink mode[32,33], where Spruit's[26] theory implies that $c = c_k$, the kink mode speed in an unstratified tube (see equation (5)). The frequency Ω is given by[26]

$$\Omega^2 = \Omega^2_{kink} = \frac{c_k^2}{4\Lambda_0^2}(\frac{1}{4} + \Lambda_0'). \quad (29)$$

The fact that both the sausage and kink modes, and also vertically propagating sound waves in a rigid tube, obey the Klein-Gordon equation means that we have a simple means of comparing the various modes, contrasting the values of the mode speed c, the associated frequency Ω, and the amplitude (through relationship (24) which governs the manner in which the amplitude of a mode grows with height z). To carry out this comparison we consider the sausage mode, kink mode and sound waves in an isothermal atmosphere, taking conditions that are typical of the upper photosphere. Table I displays the results for an isothermal atmosphere with $c_s = c_A = 7.5$ km s^{-1}, $\Lambda_0 = 125$km and $\gamma = 5/3$. The first two entries are for the sausage and kink modes in a magnetic flux tube. The other two entries are for sound waves in rigid tubes, one tube having the same exponential cross-sectional profile as the flux tube within which the sausage and kink modes propagate (namely, $A_0(z) = A_0(0)\exp(z/2\Lambda_0)$), and the other is for a vertically straight tube. The Table gives the wave speed c, the cyclic frequency $\Omega/2\pi$ (with corresponding period $2\pi/\Omega$), and the distance the wave travels before it e-folds in amplitude.

Table I A comparison of tube and acoustic waves (after Spruit and Roberts[32]).

WAVE	SPEED	FREQ $\Omega/2\pi$	PERIOD	e-FOLD
sausage mode	5.3 km s^{-1}	4.8mHz	208s	500km
kink mode	4.5 km s^{-1}	1.4mHz	700s	500km
exponential tube	7.5 km s^{-1}	4.9mHz	203s	500km
straight tube	7.5 km s^{-1}	4.8mHz	208s	250km

The main conclusion to be drawn from Table I is that while the sausage and kink modes are similar in some respects–though different from sound waves in a rigid tube–they differ strongly in their Ω-frequencies: $\Omega_{sausage} \gg \Omega_{kink}$. Thus kink modes, once generated, are likely to be readily propagated through the tubes into the upper chromosphere.[26] Sausage modes are subject to much the same fate as sound waves in the chromosphere and so may be dissipated at lower levels than the kink mode. However, direct observational evidence for the fate of tube waves is presently not available, though there is strong indirect evidence for their existence.[34]

The significance of the occurrence of the Klein-Gordon equation for a variety of tube modes (as well as vertically propagating sound waves) is that for an isothermal atmosphere the equation possesses a simple dispersion relation, namely

$$\omega^2 = k_z^2 c^2 + \Omega^2, \qquad (30)$$

where c and Ω take on values appropriate for the mode in question (be it the sausage wave, kink wave, or sound in a rigid tube). The occurrence of the Klein-Gordon equation signals the presence of a wake phenomenon first pointed out for sound waves by Lamb[35,36] and exploited for magnetic flux tubes by Rae and Roberts.[31] An impulsively generated wave will propagate along the tube with the speed c (see Table I) and trail behind its wave front a wake which oscillates with the frequency Ω. For the sausage mode this implies a wake oscillation of $\Omega/2\pi = 4.8$mHz, with a corresponding period of 208s. For the kink mode, the wake oscillates with a frequency of 1.4 mHz (period 700s).

Are spicules a consequence of the wake phenomenon in a magnetic flux tube? Like the suggestion that spicules are related to solitons, the association of the spicule with the wake phenomenon is attractive but at this stage ultimately uncertain. Hollweg[37] and Sterling and Hollweg[38] have developed a more substantial theory connecting wakes with spicules, integrating numerically the nonlinear thin tube equations for a rigid tube. While this scenario of wakes and spicules has a number of attractive features, observational evidence is ultimately required to encourage further developments along these lines.

CONCLUDING REMARKS

In this article we have given an over-view of the role of magnetic flux tubes as communication channels linking the photosphere with the atmosphere above. We have avoided detailed descriptions of any one aspect of our topic, referring the reader to some of the now extensive literature dealing with the subject. The next decade offers hope that many of the topics discussed here will see further theoretical development in the guiding light of observations such as OSL hopes to provide. The results will surely be fascinating!

ACKNOWLEDGEMENTS

It is a pleasure to acknowledge the generous support of Dr Dan Spicer in extending an invitation to me to attend the OSL Workshop and for providing partial financial support. I am also grateful to SERC for financial support to attend the meeting.

REFERENCES

1. M. Schussler, in Solar Photosphere: Structure, Convection and Magnetic Fields (IAU Symp. 138, Reidel, 1990), p. 161.
2. F. Cattaneo, OSL Workshop , these proceedings (1991).
3. G. Herbold, P. Ulmschneider, H.C. Spruit and R. Rosner, Astron. Astrophys. 145, 157 (1985).
4. A. Ferriz-Mas, Phys. Fluids 31, 2583 (1988).
5. J.H. Thomas, in Physics of Magnetic Flux Ropes (AGU Geophys. Mono. 58, Washington, 1990), p. 133.
6. B. Roberts and A. Mangeney, Mon. Not. Roy. Astron. Soc. 198, 7p (1982).
7. B. Roberts, Physics Fluids 28, 3280 (1985).
8. A.L. Molotovshchikov and M.S. Ruderman, Solar Phys. 109, 247 (1987).
9. P.M. Edwin and B. Roberts, Wave Motion 8, 151 (1986).
10. T. J. Bogdan and I. Lerche, Quart. Applied Maths. XLVI, 365 (1988).
11. E. Weisshaar, Phys. Fluids A1, 1406 (1989).
12. D.C. Braun,T.L. Duvall and B.J. Labonte, Astrophys. J. 335, 1015 (1988).
13. J. Lighthill, Waves in Fluids (Cambridge University Press, Cambridge, 1978).
14. D.D. Ryutov and M.P. Ryutova, Sov. Phys. J.E.T.P. 43, 491 (1976).
15. M.P. Ryutova, in Solar Photosphere: Structure, Convection and Magnetic Fields (IAU Symp. 138, Reidel, 1990), p. 229.
16. E. N. Parker, Cosmical Magnetic Fields (Clarendon, Oxford, 1979).
17. B. Roberts, Solar Phys. 69, 27, 39 (1981).
18. P.M. Edwin and B. Roberts, Solar Phys. 88, 179 (1983).
19. B. Roberts, in Solar System Magnetic Fields (Reidel, Dordrecht, 1985), p. 37.
20. D.J. Evans and B. Roberts, Astrophys. J. 348, 346 (1990).
21. B. Roberts, in Basic Plasma Processes in the Sun (Kluwer, Dordrecht, 1990), p. 159.
22. B. Roberts and A. R. Webb, Solar Phys. 56, 5 (1978).

23. E.N. Parker, Astrophys. J. 189, 563 (1974).
24. R.J. Defouw, Astrophys. J. 209, 266 (1976).
25. A. Ferriz-Mas, M. Schussler and V. Anton, Astron. Astrophys. 210, 425 (1990).
26. H.C. Spruit, Astron.Astrophys. 98, 155 (1981).
27. L.M.B.C. Campos, Reviews of Modern Phys. 58, 117 (1986).
28. E. G. Merzljakov and M.S. Ruderman, Solar Phys. 95, 51 (1985).
29. M.J. Ablowitz and H. Segur, Solitons and the Inverse Scattering Transform (SIAM, Philadelphia, 1981).
30. Y. Matsuno, Bilinear Transformation Method (Academic Press, New York, 1984).
31. I.C. Rae and B. Roberts, Astrophys. J. 256, 761 (1982).
32. H.C. Spruit and B. Roberts, Nature 304, 401 (1983).
33. B. Roberts, in Small-Scale Magnetic Flux Concentrations in the Solar Photosphere (Vandenhoeck & Ruprecht, Gottingen, 1986), p. 169.
34. S.K. Solanki and B. Roberts, in Solar Photosphere: Structure, Convection and Magnetic Fields (IAU Symp. 138, Reidel, 1990), p. 259.
35. H. Lamb, Proc. London Math. Soc. 7, 122 (1909).
36. H. Lamb, Hydrodynamics (Cambridge University Press, Cambridge, 1932).
37. J.V. Hollweg, Astrophys. J. 257, 345 (1982).
38. A.C. Sterling and J.V. Hollweg, Astrophys. J. 327, 950 (1988).

MAGNETIC FIELD LINE TOPOLOGY IN SOLAR ACTIVE REGIONS

N. Seehafer

Astrophysikalisches Observatorium Potsdam, O-1561 Potsdam, Germany

ABSTRACT

According to a recent study comparing magnetic fields extrapolated from photospheric measurements with chromospheric and coronal observations, the electric current helicity in active regions possesses a predominant sign in each of the two solar hemispheres. This indicates that the generation of atmospheric electric currents, which are needed as the energy source for flares and coronal heating, is a global phenomenon connected with the rotation of the Sun rather than a consequence of plasma motions within individual active regions uncorrelated between different active regions. Theoretically, the evolution of the atmospheric magnetic field may be understood as the continual distortion of an existing equilibrium by disturbances propagating upward from the photosphere and subsequent fast relaxation to a new, neighbouring equilibrium. Here current helicity proves to be an important quantity decisive for whether the (mean) magnetic field can evolve along a stable path with growing free energy. — Disruptive disturbances with an explosive release of magnetic energy, in particular flares, are generally thought to be due to current sheet formation and magnetic reconnection. Reconnection may be defined as a discontinuous change of the field line connectivity to the photosphere. Then field lines running into a magnetic null point or such that are tangential to the photosphere should play a particular role. By comparing extrapolated magnetic fields with flare observations examples have been found suggestive of a connection between tangential field lines and flare activity.

INTRODUCTION

In the theories of the solar magnetism helicities are of fundamental importance. With v, B, A and j denoting fluid velocity, magnetic induction, magnetic vector potential and electric current density, the densities per unit volume of kinetic, magnetic and current helicity are defined by

$$H_K = v \cdot \nabla \times v; \quad H_M = A \cdot B; \quad H_C = B \cdot \nabla \times B \qquad (1)$$

In this paper the particular role of current helicity is considered. It proves to be an important quantity for the build-up of currents in the solar atmosphere (and in general).

Helicities arise as a consequence of the rotation of the Sun. It is generally accepted that rotation is the cause for the global magnetic field of the Sun. The solar rotation also may be the ultimate cause for the generation of dc currents in the atmosphere. Such currents are presumably the energy source for flares and also may play a role for the non-flare heating of the atmosphere. According to a recent study[1] comparing force-free magnetic fields calculated from photospheric magnetograms with chromospheric and coronal observations, mainly in Hα and

EUV lines, H_C is predominantly negative in the northern and positive in the southern hemisphere. This dependence is presumably a result of the different actions of the Coriolis forces in the two hemispheres.

Besides the current build-up in the atmosphere, for which the helicity plays its role, in a separate section also the problem of the topological conditions for flare activity is touched upon. In particular arguments for the importance of field lines tangential to the photosphere are presented.

QUASISTATIC EVOLUTION OF THE MAGNETIC FIELD

The photosphere can to some extent be considered as a rigid wall for the superphotospheric layers of the atmosphere, since the Alfvén velocity of the medium above the photosphere is much higher than that of the deeper layers. Since, in addition, the magnetic energy dominates over all other energies, the plasma-magnetic field configuration above the photosphere evolves, except for times of explosive events, slowly through a sequence of force-free equilibrium states, characterised by the equation

$$\mathbf{\nabla} \times \mathbf{B} = \alpha_{ff} \mathbf{B}, \qquad (2)$$

where α_{ff} denotes a pseudo-scalar which in general depends on position. The current helicity and α_{ff} are related by the equation

$$H_C = \alpha_{ff} B^2. \qquad (3)$$

The slow evolution may be understood as a continual distortion of the equilibrium by disturbances propagating upward from the photosphere and subsequent fast relaxation to a new, neighbouring equilibrium. The relaxation leads to a state of minimum energy under given photospheric boundary conditions and further constraints following from the nature of the relaxation process. Without resistivity, i.e., if the magnetic field is frozen into the plasma, the relevant variational problem is

$$\int_V (\mathbf{\nabla} \times \mathbf{A})^2 dV = minimum, \qquad (4)$$

$$\delta \mathbf{A} \times \mathbf{n} = 0 \quad \text{on } \partial V, \qquad (5)$$

$$\delta \mathbf{A} = \delta \boldsymbol{\xi} \times (\mathbf{\nabla} \times \mathbf{A}) \quad \text{for some } \delta \boldsymbol{\xi} \text{ in } V, \qquad (6)$$

where $\mathbf{n}$ is the normal on the surface ∂V of the considered volume V and $\delta \boldsymbol{\xi}$ the displacement of a fluid element. Equation (6) expresses the frozen-in-field condition, Equation (5) fixes the distribution of the photospheric flux, and Equations (5) and (6) together fix the photospheric end points of magnetic field lines (the "field line connectivity"). This problem leads to the necessary condition

$$\mathbf{j} \times \mathbf{B} = 0. \qquad (7)$$

The magnetic field relaxes to a general force-free state, in which the factor α_{ff} in general is not spatially constant, and the relaxation takes place on the short Alfvén time-scale.

If there is a small amount of resistivity, the system may relax further, namely by Taylor relaxation. The corresponding variational problem is given by Equations (4) and (5) and the condition that the total magnetic helicity is conserved,

$$\int_V \boldsymbol{A} \cdot (\boldsymbol{\nabla} \times \boldsymbol{A})\, dV = const. \tag{8}$$

The final state is force-free with spatially constant α_{ff}. Taylor relaxation, involving magnetic reconnection, takes place on a reconnection time-scale, which is intermedeate between the Alfvén and diffusive time-scales.

RELAXATION AS INVERSE CASCADE

If during the slow evolution the magnetic energy is increased, then, obviously, the relaxation processes represent inverse cascades of energy. Such cascades can be described within the framework of mean-field mhd, in which it is usual to decompose all quantities into mean and fluctuating parts. Accordingly the mean value of the current helicity can be represented as the sum of the two contributions resulting from the mean and fluctuating magnetic fields, respectively,

$$<\boldsymbol{B} \cdot \boldsymbol{\nabla} \times \boldsymbol{B}> = <\boldsymbol{B}> \cdot \boldsymbol{\nabla} \times <\boldsymbol{B}> + <\boldsymbol{B'} \cdot \boldsymbol{\nabla} \times \boldsymbol{B'}>. \tag{9}$$

Angular brackets always indicate averages and a prime fluctuations.

Magnetic and velocity fluctuations lead to a mean electromotive force

$$\mathcal{E} = <\boldsymbol{v'} \times \boldsymbol{B'}>, \tag{10}$$

which for a wide range of assumptions can be written in the form

$$\mathcal{E} = \alpha <\boldsymbol{B}> - \beta \boldsymbol{\nabla} \times <\boldsymbol{B}> \tag{11}$$

with coefficients α and β which are in general tensors. The first term describes the α-effect, the second a turbulent diffusivity.

Now let the photosphere be a perfectly conducting rigid wall and V the volume above the photosphere. Then[2] under the assumptions that
(i) $<\boldsymbol{v}> = 0$,
(ii) $\boldsymbol{v'}$ describes a homogeneous, steady and isotropic turbulence,
(iii) no higher than second order correlations of $\boldsymbol{v'}$ are taken into account,
the magnetic energy in V can grow (or at least be prevented from decaying) (by the α-effect) only if there is some subvolume in which
(i) $<\boldsymbol{B'} \cdot \boldsymbol{\nabla} \times \boldsymbol{B'}> \neq 0$ and $<\boldsymbol{B}> \cdot \boldsymbol{\nabla} \times <\boldsymbol{B}> \neq 0$,
(ii) $<\boldsymbol{B'} \cdot \boldsymbol{\nabla} \times \boldsymbol{B'}>$ and $<\boldsymbol{B}> \cdot \boldsymbol{\nabla} \times <\boldsymbol{B}>$ have opposite signs,
(iii) $|<\boldsymbol{B'} \cdot \boldsymbol{\nabla} \times \boldsymbol{B'}>| > |<\boldsymbol{B}> \cdot \boldsymbol{\nabla} \times <\boldsymbol{B}>|$.
It seems interesting to note that for the inverse cascade not only the small-scale but also the large-scale field must possess current helicity (of appropriate sign), in order to be able to pick up the energy from the small scale.

To some extent the above finding can be confirmed by considering the ideal MHD stability of force-free fields (it is presenly not known whether the

α-effect survives in the limit of infinite conductivity). A necessary and sufficient condition for the linear ideal stability of a line-tied force-free field is the positive definiteness of the second-order energy variation,

$$\delta^2 W = \frac{1}{2} \int_V [(\nabla \times \delta A)^2 - \alpha_{ff} \delta A \cdot \nabla \times \delta A] \, dV \geq 0. \tag{12}$$

In some sense instability, namely the growth of a perturbation (fluctuation) at the expense of an equilibrium field (mean field) is the opposite of an inverse cascade. From Equation (12) it can be seen that instability is only possible if the helicities of the perturbation ($\delta A \cdot \nabla \times \delta A$) and of the equilibrium field (α_{ff}) have the same sign (in sufficiently extended subvolumes). (A difficulty here is that in Euation (12) not the current helicity but the magnetic helicity of the perturbation appears. There is no simple relation between these two helicities, but for important special cases, e.g. cylindrically symmetric and constant-α_{ff} force-free fields, they have the same sign.[1])

FIELD LINES TANGENTIAL TO THE PHOTOSPHERE

The magnetic field of solar active regions is typically closed, i.e. all field lines have two endpoints in the photosphere. So they define a mapping from the photosphere to itself.

Disruptive disturbances with an explosive release of magnetic energy, in particular flares, are generally thought to be due to current sheet formation and magnetic reconnection. Under the condition that the photosphere can be considered as rigid on the reconnection time-scale, reconnection may be defined as a discontinuous change of the field line connectivity to the photosphere. Now parametric representatios $x(\lambda)$ of individual field lines are obtained as solutions of the equation

$$\frac{dx}{d\lambda} = B(x). \tag{13}$$

Through any point x_0 there is a unique solution $x(x_0, \lambda)$, which, for given finite λ, is continuously differentiable with respect to the initial position x_0, provided $B(x)$ is correspondingly differentiable. If B depends on some parameter, in particular time t, then the solution $x(x_0, \lambda, t)$ is continuously differentiable also with respect to t, again provided that B is correspondingly differentiable.

Therefore there are only two possibilities for a discontinuous change of the field line connectivity: Field lines must either run into a magnetic null point (then $\lambda \to \infty$) or be tangential to the photosphere at one end.[3]

This implies in particular that reconnection within a simple loop which does not contain a null point and on the (photospheric) end faces of which the normal field component does not change its sign is impossible (such loops are "topologically stable"[4]).

Also without regard to the above definition of reconnection, the tangental field lines and/or those running into null points can be expected to play a particular role. They form surfaces of discontinuity, separatrix surfaces, for the mapping from the photosphere to the photosphere defined by the field lines. Now consider some quantity that is constant along the field lines. The superphotospheric magnetic field evolves in response to changing photospheric boundary conditions, which in turn are (little or) not influenced by the magnetic field. Then infinitesimally small changes at the photosphere can lead to a finite jump across a separatrix surface of the quantity considered.

This is most easily demonstrated for fields with an ignorable coordinate. Let $\boldsymbol{B}$ be translationally invariant in the y-direction. Then it can be written in the form

$$\boldsymbol{B} = \boldsymbol{\nabla}\psi(x,z) \times \boldsymbol{e}_y + B_y(x,z)\boldsymbol{e}_y, \qquad (14)$$

where ψ is the poloidal flux function and $\boldsymbol{e}_y$ the unit vector in the y-direction. ψ is constant along the field lines. For a force-free $\boldsymbol{B}$ then

$$B_y = f(\psi) \qquad (15)$$

with some function f, showing that B_y is constant along the field lines. Therefore photospheric shearing motions can produce discontinuities of B_y across, i.e. current sheets along the separatrices.[5,6]

Consider now a general (non-constant α_{ff}, 3D) force-free field. By taking the divergence of Equation (2) one gets

$$\boldsymbol{\nabla}\alpha_{ff} \cdot \boldsymbol{B} = 0, \qquad (16)$$

i.e., α_{ff} is constant along the field lines. Changing photospheric boundary conditions can lead to discontinuities of α_{ff} and, consequently, also of $\boldsymbol{j}$ across the separatrices. Then, though $\boldsymbol{B}$ can still be continuous (so this is a weaker discontinuity than one of the $\boldsymbol{B}$-field), the separatrices become diffusive layers, since the diffusion term in the induction equation is $\boldsymbol{\nabla} \times (\boldsymbol{j}/\sigma)$ (σ is the electrical conductivity).

These arguments are equally valid for discontinuities of the field line mapping due to tangential field lines and such due to magnetic nulls. However, the tangential field lines are much more likely to play a role in solar active regions[3]:
(i) It seems rather questionable whether magnetic nulls typically exist in active regions. E.g. for a quadrupolar configuration generated by four (magnetic) point charges of equal strength situated at equal depth below the photosphere magnetic nulls exist only if the charges are located at the corners of a rectangle.
(ii) Except for simple dipole configurations, tangential field lines are always present. The zero line of the photospheric normal field component is bridged by field lines, in part above and in part below the photosphere. Where it is bridged below the photosphere, field lines touch the photosphere from above.
(iii) By comparing extrapolated magnetic fields with flare observations examples have been found suggestive of a connection between tangential field lines and flare activity.

REFERENCES

1. N. Seehafer, Solar. Phys. <u>125</u>, 219 (1990).
2. K.-H. Rädler, N. Seehafer, in H.K. Moffatt, A. Tsinober (eds.), Topolocical Fluid Mechnics (Cambridge University Press, 1990), p. 157.
3. N. Seehafer, Solar. Phys <u>105</u>, 223 (1986).
4. N. Seehafer, Solar. Phys <u>107</u>, 73 (1987).
5. B.C. Low, R. Wolfson, Astrophys. J. <u>324</u>, 574 (1988).
6. G.E. Vekstein, E.R. Priest, T. Amari, Astron. Astrophys. 242, 429 (1991).

WEAK SOLAR MAGNETIC FIELDS

J.O. Stenflo

Institute of Astronomy, ETH-Zentrum, CH-8092 Zurich, Switzerland

ABSTRACT

Intrinsically weak magnetic fields are difficult to identify, since flux measurements (magnetograms) cannot by themselves distinguish between filling-factor and field-strength effects. The first real determinations of intrinsically weak (less than kG) fields have in fact been made only this year, using the Stokes V profiles of an infrared line pair near 1.56 μm. Many cases of discrete magnetic elements with field strengths as low as 0.4 kG have been found, immediately adjacent (within a couple of arcsec) to the normal strong-field fluxtubes that have strengths in the range 1.4–1.6 kG and a magnetic polarity that can be both the same or opposite to that of the adjacent magnetic component.

There appears to be a continuous sequence of bipolar magnetic regions of various scales, down to the spatial resolution limit, from active regions to ephemeral regions and inner-network fields. It seems likely that this sequence continues in the form of a subarcsec mixed-polarity or "turbulent" field that permeates the 99 % of the photospheric volume not occupied by the kG fluxtubes in the network. A one-sigma upper limit of 100 G to the strength of this hitherto "invisible" field has been set from line-broadening constraints, which indicates that this small-scale field is intrinsically weak. Arguments are given why the spatial spectrum of flux emergence should saturate when scales approaching the photon mean free path in the photosphere (about 100 km) are approached, which is the range of scales that may be opened up to exploration by LEST and OSL.

It is shown how correlations ("active longitudes") in the pattern of small-scale flux emergence lead to a replenishment of the global or "background" magnetic-field pattern at high heliographic latitudes in a time as short as weeks, more than two orders of magnitude faster than predicted by numerical models of the Babcock-Leighton type. There is thus a close link between the small-scale dynamics and the global solar-cycle evolution.

INTRODUCTION: EXISTENCE OF WEAK FIELDS

Since the discovery about two decades ago[1] of the highly intermittent kG structure of the sun's magnetic field, much attention has been directed towards the understanding of the physics of kG fluxtubes. It follows from the intrinsic kG nature of the magnetic flux seen in solar magnetograms and from the circumstance that the large-scale field has average flux densities of the order of 10 G or less that the typical average magnetic filling factor of the kG fields is of the order of 1 % . If however only 1 % of the photospheric volume is occupied by the strong fields, then 99 % must be occupied by something else.

In idealized theoretical descriptions of the strong fields one often uses models according to which the strong-field fluxtubes are embedded in a field-free environment. Nature however does not like zeros. The so-called "field-free" region must for reasons of finite efficiency of flux concentration mechanisms, finite magnetic diffusivity, induction effects, turbulent motions, etc., be permeated by magnetic fields filling every single volume element. The question is not if a field in this region exists, but what its properties are, and how it can be diagnosed by future powerful instruments like LEST[2] and OSL.

DIAGNOSTIC CONSIDERATIONS AND OBSERVATIONAL LIMITATIONS

A misconception that still tends to appear in the literature over and over again is the belief that when the observed magnetic flux is tiny, this suggests that the corresponding magnetic fields are also instrinsically weak. It should however be obvious from the definition of flux and field strength that a measured flux density (averaged over the spatial resolution element, whatever it might be) can only provide a lower limit to the intrinsic field strength.

Let us recall that the first determination[1] of the intrinsic kG nature of the flux outside active regions was based on observations in a very quiet region that had observed flux densities with typical values of a few G only. Using a non-linear effect of the Zeeman splitting (Zeeman saturation) to become independent of the spatial resolution, it could be shown that these tiny fluxes were actually due to kG fields, which accordingly must have had very small filling factors. The measured fluxes by themselves could provide no information whatsoever on the field strengths (apart from lower limits, which are rather useless when the filling factors are small).

Although it is true that weak fields imply small fluxes, the reverse does not hold: small fluxes do not provide information on whether the fields are weak or strong. This brings out the diagnostic problem: Weak fields cannot be diagnosed by flux measurements, we need field-strength diagnostics.

The diagnostic problem would of course be greatly alleviated by increasing the spatial resolution, as OSL and LEST want to do. With increasing resolution, the filling factor approaches unity and the lower limit approaches the intrinsic value. Another important advantage of higher resolution is that with the accompanying higher filling factor the polarization signal becomes stronger and the contrast of the structures in the magnetic-field map becomes larger. However, even with 0.1 sec of arc spatial resolution flux measurements alone do not tell us whether the filling factor is close to unity or not. The apparent flux morphology does not tell us whether we have to do with a single clump or a cluster of separate magnetic elements. Therefore, even with high-resolution instruments like LEST or OSL, flux measurements always have to be complemented by field-strength diagnostics.

Observed flux densities may be small for various reasons, and we have to have diagnostic methods to handle all these cases: (1) The fields may be intrinsically weak. (2) The filling factor may be small. (3) There may be mixed-polarity fields within the spatial resolution element, so that the net polarization effect is small. If a circular polarization signal can be recorded, the intrinsic strength of the field(s) causing this signal may be diagnosed by primarily two classes of methods (which may be combined with each other): (i) The line-ratio method (which may be generalized to a multi-line method). (ii) Direct determination of the Zeeman splitting in the infrared. If the field is so tangled that the net polarization is too small to be measured, other indirect methods have to be used to constrain the field: Zeeman broadening of unpolarized lines, Hanle-effect depolarization, and the transverse Zeeman effect (which has different symmetry properties as compared with the longitudinal Zeeman effect).[3]

Whatever diagnostic method we use, intrinsically weak fields always imply small magnetic-field induced signals to be detected. A main observational limitation is therefore the attainable signal-to-noise ratio. The observations will always be photon starved, which means that the light-collecting area of future solar telescopes is a very important property for the study of the "weak" magnetic fields that fill 99 % of the photosphere. This is one reason why the aperture of LEST was chosen to be as large as 2.4 m.

FRACTION OF FLUX IN STRONG-FIELD FORM

As we have seen, only about 1 % of the photosphere is filled with strong (kG) fields. Yet another two decade old result says that more than 90 % of the total flux that is seen in magnetograms is due to the kG fields.[4,5] This places important empirical constraints on the nature of the field, but it may also give the impression that the weak fields play an insignificant role as they seem to carry so little flux. Let us therefore take a brief look at how the determination of the fraction of strong-field flux was made, and what it really implies about the properties of the weak fields.

In its original form, the line-ratio method compares the apparent field strengths recorded simultaneously with a magnetograph in spectral lines of different Zeeman and/or temperature sensitivities. An example of a graphical representation of such measurements is given in figure 1, in the form of a scatter plot of the apparent field strengths in the two lines.[6] The reason why the points do not fall along a 45° line is that the two lines are differently affected by Zeeman saturation (the deviation from a linear relation between the measured polarization and the Zeeman splitting) and temperature line weakening. The calibration of the relation between the polarization and the magnetic field for the determination of apparent field strengths cannot take these effects into account, since they are due to "invisible", i.e., spatially unresolved structures.

By selecting a suitable line pair, the Zeeman saturation effect can be isolated from the other possible effects, which enables determinations of intrinsic field strengths in an almost model-independent way. The model dependence comes in primarily when one wants to find out to what geometrical height in the atmosphere the measured intrinsic field strength really refers.

Although the ratioing between the apparent field strengths can in principle be done for each individual point on the sun, the non-linear Zeeman saturation effect to be measured is small, and the polarization noise in the observational data is generally considerable. High accuracy can be achieved with a statistical approach, whereby the ratio is determined by a regression-line fit to the data in the scatter-plot diagram. Such an approach would not be possible if the ratio values would have a continuous spread between unity and some extreme value. Inspection of scatter-plot diagrams like that of figure 1 shows however that the points fall symmetrically along a well defined regression line, with a spread that does not seem to exceed the instrumental scatter, and with no indication of a significant population of points between the regression line and the 45° line (in the region of the diagram where the instrumental scatter does not mask this effect). The existence of a well-defined regression line indicates that the flux

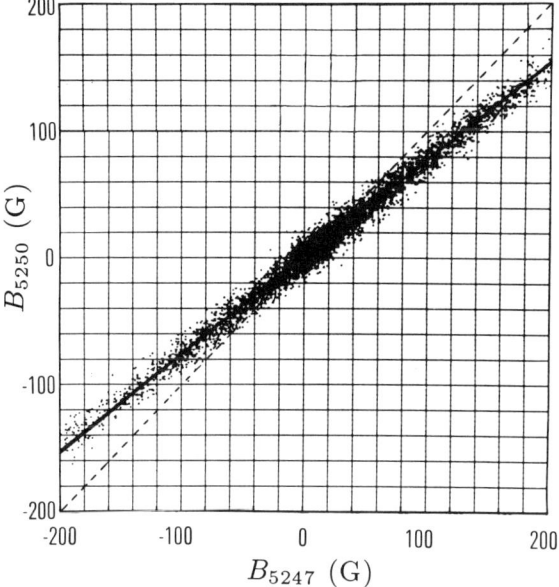

Fig. 1. Scatter plot diagram of the apparent field strengths recorded simultaneously in the Fe I 5247.06 and 5250.22 Å lines, from ref. 6. If the fields were intrinsically weak ($\lesssim 0.5$ kG), the points would fall around the 45° line.

elements have well defined properties, and thus also supports the validity of a two-component model approach for the interpretation of the data.

In a regression-type analysis (which results in kG intrinsic field strengths) it is however the points with large apparent field strengths that are given most weight, although the majority of the points are clustered around origo. As the points with small apparent field strengths are so numerous, they jointly contribute to a considerable fraction of the total flux

$$\Phi_{tot} = \sum_i |\Phi_i|, \qquad (1)$$

which represents the sum of the absolute values of the fluxes Φ_i through the spatial resolution element at position i on the sun. The question is if these apparently weak fields are also due to intrinsically kG fields, or if they are intrinsically weak. The regression-line slope does not provide an answer to this question.

To obtain a statistical answer we first define the fraction R_s of the total flux in strong-field form as

$$R_s = \sum_i |\Phi_{s,i}|/\Phi_{tot}. \qquad (2)$$

Here $\Phi_{s,i}$ is defined as the strong-field (kG) flux contribution to the flux Φ_i at disk position no. i. The remaining contribution to Φ_i is due to weak fields within the resolution element. Our task is to find the empirical value of R_s.

Let us by δ denote the factor by which the apparent field strength is affected by Zeeman saturation and temperature weakening in the kG flux elements. Assume that we for the line-ratio method have selected a line pair for which one of the lines (line no. 1) is sensitive to these effects, while the other line (no. 2) is insensitive and thus has a δ that can be assumed to be unity. In this case the slope of the regression line in the scatter-plot diagram is δ (representing line no. 1). If the "signals" (defined as the apparent field strengths) in the two lines are $S_{1,i}$ and $S_{2,i}$, respectively, at disk position no. i, the ratio between the total apparent fluxes measured in the two lines is

$$\rho = \sum_i |S_{1,i}| / \sum_i |S_{2,i}|. \qquad (3)$$

If all fluxes were due to strong fields, i.e., $R_s = 1$, then we would expect to have $\rho = \delta$. If on the other hand R_s were ≈ 0, then ρ would be ≈ 1. In general we have approximately[7]

$$R_s \approx \frac{1-\rho}{1-\delta}. \qquad (4)$$

As δ and ρ can both be obtained from the scatter-plot data, we can determine what R_s is. The observations show that $R_s \approx 1$, with an uncertainty due to instrumental scatter that corresponds to 10 % in R_s. From this follows the conclusion that at least 90 % of the total flux seen by the magnetograph is in strong-field (kG) form. For more details, see refs. 4, 5, and 7.

In this conclusion about the 90 % strong-field flux, the qualification "seen by the magnetograph" is very essential. It means that when the flux distribution over the solar disk is smeared by a spatial window that has the same size as the effective spatial resolution element of the magnetograph observations, then at least 90 % of this flux comes from kG sources. This is a resolution-dependent conclusion, and it does not tell us what the fraction of strong-field flux would be for a fully resolved sun. The smallest sampling aperture of the observations used to derive the 90 % limit was 2.4" × 2.4", and due to seeing effects the effective area was somewhat larger. What is not constrained is the possibility that on a scale smaller than a few sec of arc the photosphere is permeated by a field of mixed polarities, such that the polarization contributions of the opposite-polarity fluxes cancel when averaged over a few sec of arc. This mixed-polarity field need in principle not be weak, since it is not seen in the magnetograms, and it may carry much more flux than the 10 % of the total magnetic fluxes.

The most direct way to determine the properties of this hitherto "unseen" field would be to increase the spatial resolution, e.g. with OSL and LEST, so that what is presently unseen becomes seen. The problem is to preserve high polarimetric accuracy with the increased resolution. This always leads to photon-starved observations and thus to the demand for the largest possible telescope aperture. In the meantime, however, several indirect methods can be applied to further constrain the properties of the mixed-polarity field. We will return to this issue later below, but first discuss some new results from infrared diagnostics, which reveal the existence of discrete, weak-field fluxtubes (without contradicting the approximate 10 % weak flux limit).

"WEAK" FIELDS FROM INFRARED DIAGNOSTICS

The near infrared region around 1.6 μm has a unique potential for magnetic field diagnostics, for the following main reasons:
- The ratio between the Zeeman splitting and the line width is about three times larger than in the visible.
- The opacity minimum is in this wavelength region, which means that we see deeper in the atmosphere (and therefore also to higher field strengths) than in any other part of the solar spectrum.
- This spectral region is readily accessible to fairly standard polarimetric techniques (polarimetry in the recently explored 12 μm region[8,9] is also of great interest but technically much more difficult).

Although for a completely Zeeman-split line the field strength can be measured directly from the separation between the σ components,[10] we also in this case need a line-ratio method to sort out the physics. The main reason is that the line is not only split, but the σ components are also broadened in excess of what can be accounted for by the Doppler effect — there is Zeeman broadening due to a distribution of field strengths.[11] To separate how much broadening is due to the Zeeman effect and how much is due to the Doppler effect, one needs to combine the simultaneously recorded polarimetric information in two similar lines of different Zeeman sensitivity.[12]

Recent recordings of Stokes I (intensity) and V (circular polarization) profiles of the Fe I line pair at 1.5648 μm (the "Zeeman line", with a Landé factor of $g = 3.0$) and 1.5653 μm (the "Doppler line", with $g = 1.5$) at a large number of disk positions, using the NSO MacMath grating spectrometer with a sampling aperture of $3'' \times 3''$, have revealed a number of cases of strange-looking or "pathological" Stokes V profiles[13,14] that look very different from what has ever been seen in the visible, with the exception of some recent observations of peculiar 6302 Å Stokes V profiles across a flaring neutral line.[15] The most "pathological" case analysed so far is shown in figure 2. It is clear that two-component models, with one magnetic and one non-magnetic component, are unable to reproduce profiles like those of figure 2. The question is how many discrete magnetic components are actually needed.

It may seem that with one or a few additional model components we get so many free model parameters that almost anything could be fitted. This is however not at all so. If we for instance try to fit the most "normal-looking", single-peaked infrared Stokes V profiles with synthetic profiles derived from a model in which the field is assumed to be height independent, then a whole (quasi-continuous) distribution of many components of various field strengths would be needed to account for the observed Zeeman broadening (as we need to reproduce the widths of the "Zeeman" and "Doppler" lines simultaneously).

In a realistic and selfconsistent model of the magnetic elements, however, the field strength has to decrease with height to ensure horizontal pressure balance as the outside gas pressure drops exponentially with a small scale height (about 150 km). In the deeper photospheric layers that we are concerned with in the infrared, the simple thin-tube approximation is fully adequate to describe

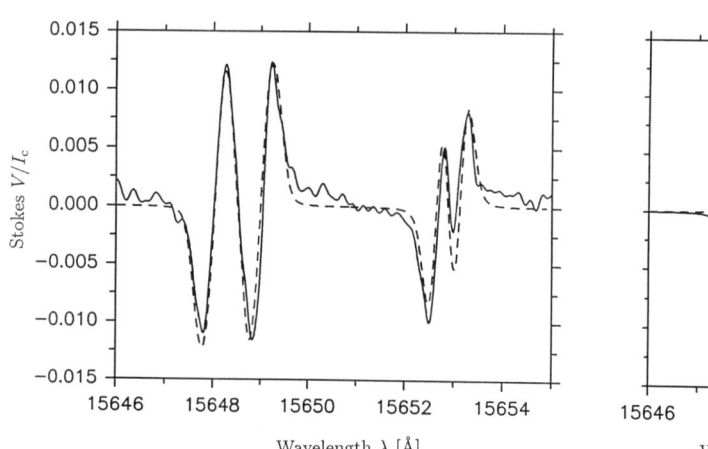
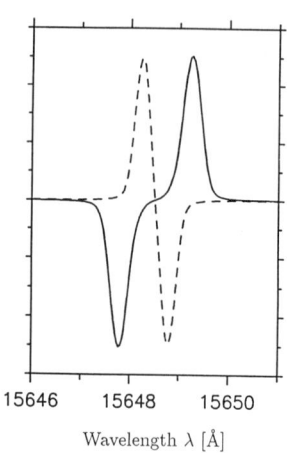

Fig. 2. Abnormal-looking, double-peaked Stokes V profiles of the infrared Fe I lines at 1.5648 and 1.5653 μm, from ref. 14. The solid line in the left diagram is the observed Stokes V (normalized to the continuum intensity I_c) spectrum, the dashed line is the synthetic spectrum produced by a model with two discrete magnetic components and one non-magnetic component. The field strengths of the two magnetic components are 1.70 and -1.05 kG, respectively, at zero geometrical height. Their separate Stokes V contributions are shown by the solid and dashed curves of the right diagram.

the magnetohydrostatic fluxtube properties. When such a selfconsistent fluxtube model is used to calculate synthetic Stokes spectra, the field-strength variation along the line of sight for a single, discrete magnetic component is sufficient to account for the Zeeman broadening of the normal, single-peaked Stokes profiles. In the case of abnormal-looking profiles, even for the most "pathological" cases like that of figure 2, two discrete magnetic components are sufficient to achieve an adequate fit (as seen by comparing the solid and dashed curves in the left diagram of figure 2). In none of the so far analysed 27 spectra was a third magnetic component needed. This is a highly non-trivial and intriguing result, which lends support to the concept of discrete fluxtubes.

The individual Stokes V profiles needed to fit the 1.5648 μm line in figure 2 are shown in the right diagram. Their field strengths at zero geometrical height are 1.70 and -1.05 kG. This means that we have two discrete flux elements of opposite polarities and significantly different field strengths within a distance of less than a few arcsec from each other. Other observed spectra require for a fit one strong and one weak (well below kG) magnetic component adjacent to each other, and these two components may in some cases have the same, in other cases opposite polarities.

Figure 3 gives a summary of the field strengths determined from the 27 analysed spectra, as a function of the Stokes V area for a given magnetic component (defined as the average of the blue and red wing areas of V/I_c, expressed in mÅ — it is approximately proportional to the magnetic filling factor). The open circles represent the cases that only require a single magnetic component, the pluses the cases that require two magnetic components (there are thus two pluses for each such spectrum, whereas there is only one circle per spectrum for the one-component case). A regression line has been drawn through the points above 1.3 kG, to indicate more clearly the typical field-strength level and the slight dependence on magnetic filling factor for the "normal" kG flux elements.

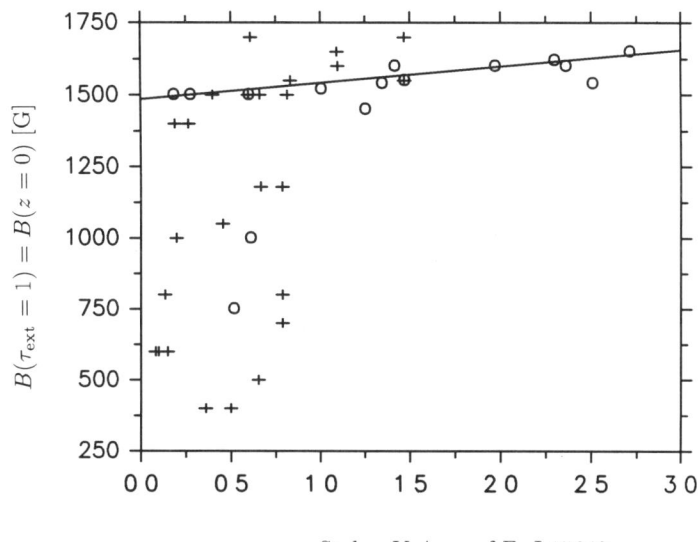

Fig. 3. Field strengths (at geometrical height zero, i.e., where the external optical depth at 5000 Å is unity) determined from 27 infrared Stokes V spectra, vs. the area (average of blue and red line-wing areas) of the synthetic Stokes V/I_c profile of the Fe I 1.5648 μm line, in mÅ. The open circles represent spectra that could be fit using a single magnetic component, while the pluses represent each magnetic component for the spectra where two magnetic components were required for a fit. From ref. 14.

With the exception of two deviating cases with significantly weaker fields, the isolated magnetic components represented by the circles have field strengths at zero geometrical height in the range 1.4–1.6 kG. For the cases requiring two magnetic components, the stronger-field component has field strengths in the same range as the isolated components, whereas the field strengths of the weaker-field components are spread over a large interval, down to values of about 0.4 kG. This work[14] represents the first identification of intrinsically weak flux elements. The weakest fields (0.4 kG) have been found adjacent to strong-field components, while the weakest isolated element found so far has a strength of 0.75 kG.

The model fits show that the elements with strengths in the range 1.4–1.6 kG have a plasma β of about 0.3, which is a factor of six smaller than the value of 1.8 that represents the theoretical limit beyond which the fluxtubes are convectively stable.[16] These fluxtubes are thus highly evacuated. For the weaker-field elements the plasma β is naturally larger.

If we sum up the flux contributions from the components in figure 3 with field strengths below 1 kG, they represent about 10 % of the total flux, which is consistent with our previous strong-field flux constraint. It is however not clear how representative the present sample of 27 points on the sun is, so it would be important to explore the field-strength and flux distributions for a more complete statistical sample, to try to determine what the typical distributions of intrinsic field strengths in different regions of the solar disk are.

Finally we note that synthetic Stokes V profiles calculated from the models with two magnetic components needed to fit the "pathological" infrared profiles, are in the visible spectral region all single peaked and look quite "normal", although these cases give a 5250–5247 magnetic line ratio that shows large scatter around the "nominal" value for the line ratio. Due to the much larger Zeeman splitting the infrared lines can reveal new, "hidden" magnetic structures, and allow the determination of fields that are intrinsically much weaker than can be measured in the visible. The 1.6 μm region thus has considerable future diagnostic potential to be exploited. Combined with the line ratio in the visible, the infrared observations can also be used to diagnose the height variations and the temperature-opacity structure of the fluxtube interiors.

MIXED-POLARITY UNRESOLVED FIELDS

As a mixed-polarity field carries no net flux when averaging over an area larger than the mixing scale, indirect methods have to be applied to constrain the field. A simple parametrization to characterize this "invisible" field is then needed. We will assume a field of strength B with a typical mixing scale (separation between the opposite polarities) d and an angular distribution of the field vectors given by $\cos^a \theta$, where θ is the angle between the field and the vertical direction. The distribution is isotropic for $a = 0$, and becomes exclusively vertical when $a \to \infty$. In addition we should add a magnetic filling factor α. These parameters are likely to depend on each other. In particular should a and α depend on B: For large (kG) values of B, we expect small values of α (since otherwise the Zeeman broadening of unpolarized spectral lines would be excessive) and large values of a (since the buoyancy force in the vertical direction becomes large). We will however focus our attention on the weak fields expected to permeate the atmosphere at the small-scale end of the spatial spectrum, and which thus should have a filling factor of approximately unity. It is the invisible field in empty-looking quiet regions of magnetograms that we are after.

Some of the present observational constraints are summarized in figure 4.[17,18] The approximate upper limit to the mixing scale d (horizontal axis) is indicated by the slanted line marked "limit from magnetograms".[10,19] If the polarity separation were larger, then the best high-resolution magnetograms would begin to resolve the field, and it would no longer be "invisible". The indicated limit is approximate of course, since the large-scale end of the mixed-polarity field may be identified with the already partially resolved inner-network fields (seen with an effective resolution of a couple of arcsec) in the center of the supergranular cells.[20,21] However, this field appears intermittent with a probably small but still undetermined filling factor, and what concerns us here is the field in the regions between the observed flux patches. It is not known whether the inner-network fields are intrinsically strong (small filling factors) or weak.

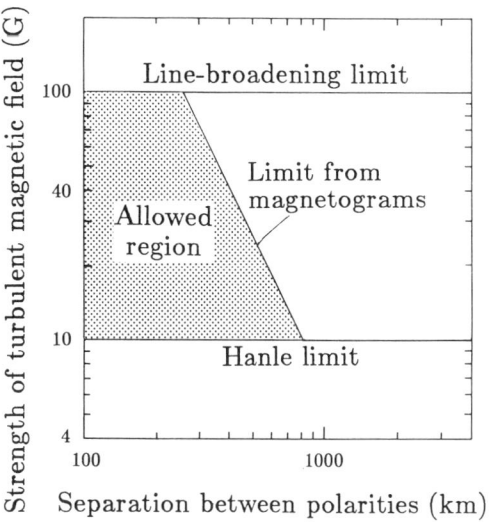

Fig. 4. Observational constraints indicating some properties of the hitherto "invisible" mixed-polarity or "turbulent" magnetic field, from ref. 17. The shaded region that remains unconstrained at present should become accessible with LEST and OSL down to the 100 km scale.

An upper limit of 100 G to the strength B of a space-filling field comes from a search for a possible correlation between the Landé factor and the width of unpolarized line profiles.[22] A field of arbitrary direction, even when there is no net flux, should contribute to a Zeeman line broadening that is proportional to the Landé factor. From the absence of such a correlation an upper limit to B could be determined. The 100 G is a 1-sigma limit. The 3-sigma limit to the Zeeman broadening corresponds to a limit on B that is larger by $\sqrt{3}$. The limit refers to spatially unresolved observations and does not provide any constraint on spatial scales.

A tentative lower limit to B of 10 G is suggested by estimates of the Hanle-effect depolarization needed to explain the lower than expected linear polarization caused by radiative scattering in the cores of resonance lines observed near the solar limb.[23] Since however the cores of these strong lines near the limb are formed near the temperature minimum or even higher, the Hanle-effect depolarization may be affected by the extended canopies of network fields,[24] and thus may not relate to the photospheric mixed-polarity fields that we are concerned with. The lower limit therefore might not apply, so the "allowed region" may be less firmly constrained than suggested by figure 4. Note also that there is no lower limit to the scale sizes — the shaded region should continue indefinitely to the left.

Another constraint on the field not shown in figure 4 can be obtained from linear-polarization observations of the transverse Zeeman effect. The transverse Zeeman effect has different symmetry properties as compared with the longitudinal Zeeman effect: the sign of Stokes Q does not change when the field direction is reversed. Therefore the cancellation effects for a mixed-polarity field are different — for a field with an isotropic distribution of field vectors Stokes Q vanishes, but for a vertical field of mixed polarities at some distance from the center of the solar disk, the Stokes Q signal survives the averaging. Observations of Stokes Q (in combination with Stokes V data) may thus provide a constraint on the angular distribution parameter a. Using such arguments, an upper limit to a of 1.3 has been derived,[3] which means that the angular distribution cannot be very far from isotropic.

It is hard to push the described indirect methods much farther. To explore the mixed-polarity field we urgently need increased spatial resolution coupled with high polarimetric accuracy, to push the magnetograph limit all the way to the left edge (100 km) of the diagram, as is planned with LEST and can be done with reduced polarimetric accuracy with OSL.

Although the net fluxes seen by magnetographs with moderate resolution are predominantly due to instrinsically strong (kG) fields, discrete fluxtubes cannot remain kG down to any scale size. A fundamental physical scale near which a definite break should occur is the horizontal photon mean free path in the photosphere, which is about 100 km and below which a structure can no longer be thermally insulated from the surroundings (as required for the mechanism of convective collapse of fluxtubes, for instance). This scale is much larger than the ohmic diffusion scale and may be resolved, at least partially, by OSL and LEST. At the small-scale end the fields must be weaker, but we do not know the spatial spectrum of the subarcsec flux. MHD theories of turbulence generally predict equipartition between the kinetic and magnetic energies at the small-scale end of the spectrum, but we do not know at what scales this equipartition is reached for the gravitationally stratified solar atmosphere.

Ordinary turbulence theory may in fact be out of place to describe what really happens in the complex solar atmosphere. When we speak of "mixed-polarity fields", we are physically dealing with a number of bipolar magnetic elements that emerge, separate, fragment, collide and annihilate, etc. At intermediate scales the partially resolved inner-network field[20,21] might be regarded as a large-scale portion of the spectrum of mixed-polarity fields, which at still larger scales has its counterpart in ephemeral active regions, and which ultimately, at the largest scales, is represented by the active regions.

ORIGIN AND GLOBAL ROLE OF THE MIXED-POLARITY FIELDS

A mixed-polarity magnetic field consists of many bipolar magnetic elements coming and going. Resolved bipolar fields are observed at all scales on the sun, and in fact there seems to be an apparently continuous sequence of emerging bipolar regions, from the largest scales in the form of active regions (AR), via intermediate scales in the form of ephemeral active regions (ER), to the smallest presently resolved scales, in the form of inner-network fields (IN). There is no reason to expect that this sequence should end with IN merely because our resolution ends there, but there is probably a continuation to the mixed-polarity fields discussed in the preceding section, which we may call "turbulent" fields (TF).

While AR occur only at low heliographic latitudes, ER have a substantially wider latitude distribution,[25] and IN may be almost latitude independent, although available data on the IN latitude variation are scarce.[20,21] In line with this sequence, we expect TF to be ubiquitous all over the sun.

The observed flux emergence rates of AR, ER, and IN are in the proportions 1:100:10,000,[21] while their spatial scales are in the approximate proportions 25:5:1. Let $d\Phi/dt$ be the overall flux emergence rate, where Φ represents the absolute flux integrated over the whole solar surface, i.e., $\int |B|\, dS$. It thus represents the accumulated or global effect of many local flux emergence events. While $d\Phi/dt$ is approximately 10^{20} Mx/day for AR, it is about 10^{22} for ER and 10^{24} Mx/day for IN.[21] To express these values as functions of size, we need similar estimates for d. Although there seems to be a continuous size distribution of bipolar elements, the typical values of d for the three categories of emerging regions are 3″ for IN, 15″ for ER, and 75″ for AR. With these estimates, we obtain[26]

$$\frac{d\bar{B}}{dt} = 350(d'')^{-2.8} \text{ G/days}, \qquad (5)$$

where

$$\frac{d\bar{B}}{dt} = \frac{1}{4\pi r_\odot^2} \frac{d\Phi}{dt} \qquad (6)$$

represents the flux emergence rate in terms of a globally averaged field strength $\bar{B}$. The size d is given in sec of arc.

Although Eq. (5) summarizes in condensed form crude estimates only, it gives a hint of the relative importance of various spatial scales. Because of the rather large negative exponent (−2.8) for d, the smallest scales dominate the build-up of a global $\bar{B}$. Inserting the value of d for AR, we get $d\bar{B}/dt = 8$ G/11 yr, within the range expected from the Babcock-Leighton model of the solar cycle.[27−29] For IN, the emergence rate is however 16 G per day, which is four orders of magnitude larger in terms of the globally averaged flux. If we would extrapolate down to a scale of 0.1″ (the approximate resolution limit of OSL and LEST), we would obtain 150 G per min! The atmosphere would quickly "choke" from so much small-scale flux unless the flux removal rate is as fast as the emergence rate, but there are physical limits to the removal rate. To avoid divergence, the trend given by Eq. (5) has to be broken or have a cut-off at small scales, and it is likely that this fundamental change takes place in the size range accessible to OSL and LEST.

A typical value for the globally averaged background field is 10 G. If we let τ_{10} represent the time over which a global field of $\bar{B} = 10$ G is built up by flux emergence (before it is removed), Eq. (5) gives

$$\tau_{10} = 0.03(d'')^{2.8} \text{ days}. \qquad (7)$$

This expression again demonstrates how the build-up time gets shorter with decreasing scale size.

The circumstance that the flux emergence rate at small scales is orders of magnitude larger than that of AR however does not by itself imply that the accumulated effect of the many small events dominates the large-scale evolution. If the flux emergence occurs randomly, then only a random background pattern would be produced, which would not contribute to the solar-cycle evolution of the large-scale polarity patterns. If on the other hand there are tiny but systematic large-scale correlations in the pattern of small-scale flux emergence, e.g. if some kind of "active longitudes" (a concept that in the past[30,31] has been used for active regions only) are preferred in the emergence of small-scale flux, then the rapid turn-over rates obtained for small values of d will apply to the global field patterns that are generated in this way.

Evidence for the existence of such "active longitudes" and for a turn-over time of the global field pattern at high heliographic latitudes as short as weeks (more than two orders of magnitude shorter than the turn-over time in the Babcock-Leighton model) has come from careful correlation analyses of the evolution of the large-scale field pattern. The correlation methods applied have been of two different kinds, yielding complementary information: (1) The longitude displacement method,[32] and (2) the pattern recurrence method.[33,34]

The longitude displacement method cross-correlates the pattern of magnetic fields in consecutive magnetograms, thereby determining the "proper motion" or phase velocity of the pattern. The derived phase velocity displays a steep differential rotation law[32] that agrees with the law determined from the measurements of Doppler shifts[35] (which gives the plasma velocity).

Such an agreement between the plasma and phase velocities contradicts the predictions of the Babcock-Leighton model. Detailed numerical modelling of the various processes in the Babcock-Leighton model (shearing by differential rotation, surface random walk by turbulent diffusion, meridional circulation) has shown that for all parameter choices the pattern develops a rotational phase velocity that increasingly deviates from the plasma velocity and becomes quasi-rigid.[36] Any model of the Babcock-Leighton type, in which all flux emergence takes place in the form of low-latitude active regions, and for which the turn-over time of the global field is of the order of 11 yr, gives rotational properties utterly different from those observed for the real sun. When the longitude displacement method is applied to simulated data from such models, the observed steep differential rotation law is not retrieved, but instead a quasi-rigid one.

The circumstance that the observed phase velocity agrees so well with the plasma velocity is evidence for a turn-over time scale of the high-latitude "background" fields that is much shorter than years, which means that these fields cannot derive their fluxes from the dispersal of active regions. A much tighter upper limit to the global turn-over time is obtained from the second correlation method, the pattern recurrence method, with which the magnetic fields sampled daily around the central meridian are used for an autocorrelation analysis, to determine the period of recurrence of the pattern to the central meridian after an integer number of solar rotations. This method results in a quasi-rigid rotation law, which is almost independent of whether a lag of 1, 2, 3, or 4 rotation periods are used to determine the rotation rate.[33,34]

If the turn-over time scale for the high-latitude fields were longer than the synodic rotation period (about 27 days), then the recurrence method would give almost the same rotation rate as the longitude displacement method, i.e., a steep differential rotation law. The results of the recurrence method thus show that the turn-over time has to be smaller than 27 days, but it should be longer than

4 days, since this was the maximum lag that could be used with the longitude displacement method (and which still gave a steep differential rotation law). According to Eq. (7), such a time scale of a couple of weeks corresponds to the build-up time of the global field by bipolar magnetic elements in the size range around 10 sec of arc. This size range is intermediate between the IN and ER scales and is accessible to observations. Available observations[20] actually hint at a turn-over time of this order for the overall flux.

The above example demonstrates the close link between small-scale field dynamics and the evolution of the global magnetic field and the solar cycle. Much of the fundamental physics occurs in the small-scale portion of the sequence AR–ER–IN–TF, which extends into the hitherto unknown domain beyond the presently available spatial resolution. LEST and OSL will be able to open the door to this unexplored territory. The scale where the small-scale emergence rate will saturate has not yet been observed but should be accessible to OSL and LEST, which should also be able to nearly resolve the fundamental length scale corresponding to the horizontal photon mean free path in the photosphere. These scales may hold the key to a unified understanding of both the small-scale and the global aspects of solar MHD.

ACKNOWLEDGEMENT

I am grateful to Sami Solanki for helpful comments on the manuscript.

REFERENCES

1. J.O. Stenflo, Solar Phys. 32, 41 (1973).
2. A.A. Wyller (ed.), LEST (The Royal Swedish Acad. Sci., Stockholm, 1991).
3. J.O. Stenflo, Solar Phys. 114, 1 (1987).
4. R. Howard, J.O. Stenflo, Solar Phys. 22, 402 (1972).
5. E.N. Frazier, J.O. Stenflo, Solar Phys. 27, 330 (1972).
6. E.N. Frazier, J.O. Stenflo, Astron. Astrophys. 70, 789 (1978).
7. J.O. Stenflo, IAU Symp. 71, 69 (1976).
8. D. Deming, R.J. Boyle, D.E. Jennings, G. Wiedemann, Astrophys. J. 333, 978 (1988).
9. D. Deming, T. Hewagama, D.E. Jennings, G. Wiedemann, in L.J. November (ed.), Solar Polarimetry (NSO, Sunspot, NM, 1991), pp. 341–355.
10. J.W. Harvey, Highlights of Astronomy 4, 223 (1977).
11. J.O. Stenflo, S.K. Solanki, J.W. Harvey, Astron. Astrophys. 173, 167 (1987).
12. I. Zayer, S.K. Solanki, J.O. Stenflo, Astron. Astrophys. 211, 463 (1989).
13. W. Livingston, in L.J. November (ed.), Solar Polarimetry (NSO, Sunspot, NM, 1991), pp. 356–360.
14. I. Rüedi, S.K. Solanki, W.C. Livingston, J.O. Stenflo, Astron. Astrophys., submitted.
15. A. Skumanich, B. Lites, in L.J. November (ed.), Solar Polarimetry (NSO, Sunspot, NM, 1991), pp. 307–317.
16. H.C. Spruit, E.G. Zweibel, Solar Phys. 62, 15 (1979).
17. J.O. Stenflo, Adv. Space Res. 4, 5 (1984).

18. J.O. Stenflo, Astron. Astrophys. Rev. 1, 3 (1989).
19. T.D. Tarbell, A.M. Title, S.A. Schoolman, Astrophys. J. 229, 387 (1979).
20. S.F. Martin, IAU Symp. 138, 129 (1990).
21. H. Zirin, Solar Phys. 110, 101 (1987).
22. J.O. Stenflo, L. Lindegren, Astron. Astrophys. 59, 367 (1977).
23. J.O. Stenflo, Solar Phys. 80, 209 (1982).
24. M. Faurobert-Scholl, Astron. Astrophys., in press.
25. K.L. Harvey, J.W. Harvey, S.F. Martin, Solar Phys. 40, 87 (1975).
26. J.O. Stenflo, in A. von Alvensleben (ed.), JOSO Annual Rep. 1990 (Kiepenheuer Inst., Freiburg, Germany, 1991), pp. 49–56.
27. H.W. Babcock, Astrophys. J. 133, 572 (1961).
28. R.B. Leighton, Astrophys. J. 140, 1547 (1964).
29. R.B. Leighton, Astrophys. J. 156, 1 (1969).
30. C. Sawyer, Ann. Rev. Astron. Astrophys. 6, 115 (1968).
31. R.S. Bogart, Solar Phys. 76, 155 (1982).
32. H.B. Snodgrass, Astrophys. J. 270, 280 (1983).
33. J.O. Stenflo, Astron. Astrophys. 210, 403 (1989).
34. J.O. Stenflo, Astron. Astrophys. 233, 220 (1990).
35. R. Howard, J.M. Adkins, J.E. Boyden, T.A. Cragg, T.S. Gregory, B.J. LaBonte, S.P. Padilla, L. Webster, Solar Phys. 83, 321 (1983).
36. N.R. Sheeley, Jr., A.G. Nash, Y.-M. Wang, Astrophys. J. 319, 481 (1987).

MAGNETIC FIELDS, OSCILLATIONS, AND HEATING IN THE QUIET SUN TEMPERATURE MINIMUM REGION

John W. Cook
E.O. Hulburt Center for Space Research
Naval Research Laboratory, Washington, DC 20375 USA

ABSTRACT

The High Resolution Telescope and Spectrograph (HRTS) instrument contains a broadband spectroheliograph which has been tuned on sounding rocket flights to cover a passband centered on 1600Å, where the predominant flux contributor is continuum emission from the temperature minimum region (approximately 70% of the integrated intensity in quiet regions). I discuss the HRTS observations of the temperature minimum region in quiet areas and their relation with magnetic fields, 5 minute oscillations, and heating. The brightness temperature of solar fine structure elements composing the supergranular network is found to be linearly proportional to the local absolute value of magnetic field strength. In cell centers, there is evidence for a 250 s period oscillation occuring in 10 arc sec scale patches, which however is energetically unimportant to the local heating budget. I discuss an interpretation in which a basal heating and 5 minute type oscillations occur globally, while the network bright points occur in magnetic regions, heated perhaps from partial dissipation of Alfven waves (whose energy flux is linearly proportional to B) in individual elemental 1500 G (at the photosphere) flux tubes which expand to form the temperature minimum fine structure bright points.

SOLAR ATMOSPHERIC STRUCTURE OF QUIET REGIONS

The central problem of the physics of the solar atmosphere is to understand the physical processes which reverse the temperature decline in the upper photosphere, as consistent with radiative equilibrium, and instead lead to the temperature rise from the temperature minimum at 4400 K, through the chromosphere and transition region, up to the several times 10^6 K corona. This input of additional energy is clearly connected with the magnetic field, either as a channelling agency, such as for magneto-acoustic waves, or more directly through input of energy from changes in the magnetic field such as reconnection.

The observed structure of the solar photosphere is a granular pattern with individual granules of order 1–2 arc sec in size which rise and fall vertically, have sideways motions of order 1 km s^{-1}, and last around 8 minutes. But by the temperature minimum, where the atmosphere already departs from a radiative equilibrium model and additional heating is occuring,[1] the observed structure in quiet areas is the supergranular network.

Small network elements composed of clumped arc sec scale bright points occur at edges of the supergranulation, coincident with stronger field regions in photospheric magnetograms, while the cell centers are filled with the order of 20 individual bright points evolving on a 1 minute timescale.[3] Figure 1 illustrates a Ca II image, HRTS 1600Å spectroheliogram, and photospheric magnetogram

56 Magnetic Fields, Oscillations, and Heating

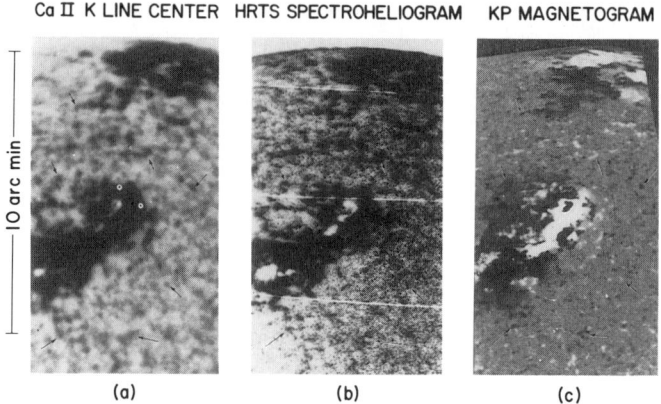

Fig. 1. Comparison of Sac Peak Ca II image, HRTS 1600Å spectroheliogram, and Kitt Peak magnetogram from the day of the HRTS II flight.

from the day of the HRTS II flight on 13 February 1978, showing the similarity of structure in Ca II and the 1600Å image, and the relationship of bright areas (dark in these negative images) with strong field regions in the magnetogram.[3] I will show that the brightness of the network is a linear function of the absolute magnitude of the magnetic flux; even magnetic bipoles, unlike in the corona, are no brighter than implied by their absolute flux strength.[4]

This supergranular network structure of quiet regions can be observed throughout the chromosphere and into the transition region, with a fair degree of correspondence in location of network elements in images from a range of temperatures. This basic network structure persists up to at least 500,000 K, as can be seen in Skylab data from the NRL SO82A instrument.[2,8] Images in Ne VII, formed at 500,000 K, show that this temperature represents something of a transitional range for solar morphology. The quiet disk is still organized in the network pattern, but loop structures are beginning to appear in active regions. At temperatures of 10^6 K (seen in Mg IX) and higher (Fe XV), the disk and limb show large complete loop systems, typically connecting separated active regions, and the supergranular pattern has disappeared. The basic quiet Sun structure up to at least 500,000 K is closely related to structures of lower temperature plasmas, instead of to the corona.

RELATIONSHIP OF BRIGHTNESS AND MAGNETIC FLUX

The High Resolution Telescope and Spectrograph (HRTS) instrument consists of a 30 cm diameter telescope; a broadband spectroheliograph which has been tuned in different flights to either 1600Å to view the temperature minimum (in quiet areas approximately 70% of the total flux in the passband is continuum emission, arising from the T_{min} region), or to 1550Å to view limb structures in C IV 1548Å and 1550Å; a slit spectrograph which can cover a wavelength range from 1175–1710Å with 0.05Å spectral resolution; and an Hα system using a temperature controlled Fabry–Perot filter, which can both display images using a TV camera and record them on film. Both the spectroheliograph and Hα systems use reflected images from the spectrograph slit jaw mirrors. Slit spectra and spectroheliograph images are recorded on film. The spectrograph slit is 920 arc sec, or almost a solar diameter, in length, and the spatial resolution is potentially an arc second or better.

Using HRTS 1600Å spectroheliograms and Kitt Peak magnetograms, Cook and Ewing[4] examined the quantitative relationship of brightness temperature at 1600Å to photospheric magnetic field strength in the quiet Sun. We used a technique which obtained the best-fit relationship of a given functional form between two histogram distributions, in our case for brightness temperature at 1600Å and for the distribution of absolute magnetic flux, in a 486x452 arc sec sample quiet area from the HRTS V flight on 11 December 1987.

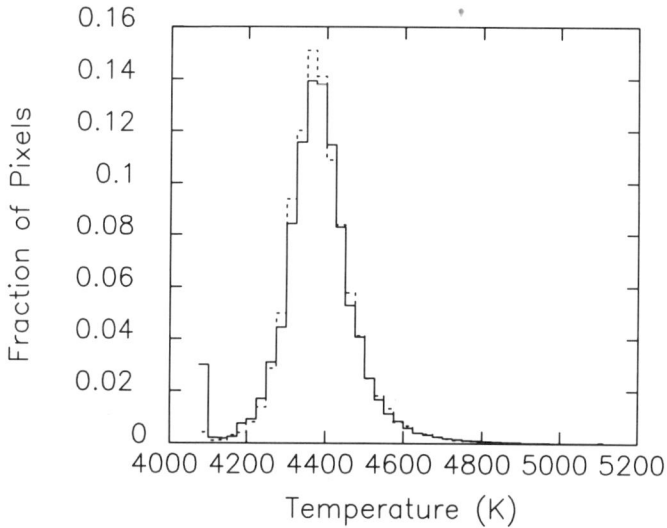

Fig. 2. Histogram distribution of brightness temperature in a sample quiet area.

Figure 2 illustrates the distribution of brightness temperature at 1600Å which we found for our sample quiet area (the solid and dashed lines were obtained from two different exposures of the same sample area). The functional form used for mapping one distribution to the other is very general, and was originally meant to reproduce the S curve shape of film characteristic curves in photometric photometry.[6] The observations alone determined the basically linear fit, which was not imposed at the start. In Figure 3 we show the functional relationship which we found (the solid and dashed lines from two different 1600Å exposures of the same sample area are again in good agreement).

We estimated the additional energy flux necessary to maintain the network bright points by integrating the LTE approximation $\Delta E = 16\sigma T_0^3 \Delta T \tau$, where ΔT is the observed brightness temperature excess in network bright points above

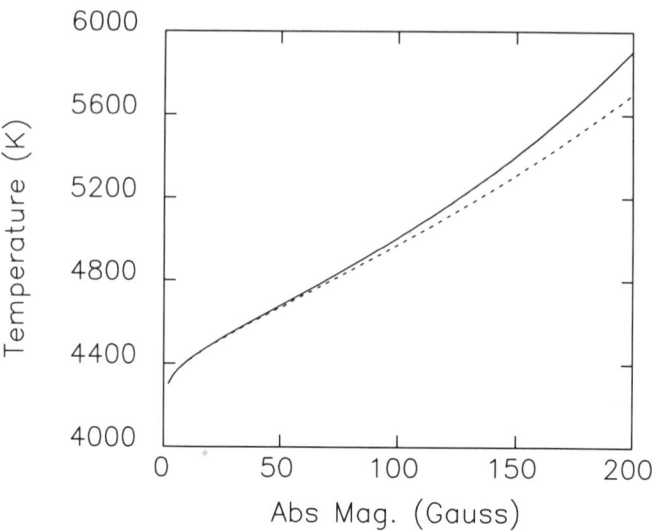

Fig. 3. Relationship of 1600Å brightness temperature vs. absolute magnetic field strength in a sample quiet area.

an average temperature T_0, and τ is the optical depth (5000Å) at the temperature minimum level, over the observed distribution of brightness temperature above 4600 K, which represents the bright point population. This average additional energy flux, 7.5x10^6 ergs cm^{-2} s^{-1}, is directly comparable to the flux required to heat the normal average global background at the T_{min} level, 8x10^6 ergs cm^{-2} s^{-1}, from the prescribed heating models of Anderson and Athay.[1]

This linear result is consistent with heating by Alfven waves, whose flux $F_A = \rho v^2 V_A$, where $V_A = B/(4\pi\rho)^{1/2}$ is the Alfven velocity, is linear in B. We estimated the potential flux at photospheric levels produced from granular buffeting in magnetized areas. Using v~1 km s^{-1} as the typical horizontal granular velocity[10] and $\rho = 2.7 \times 10^{-7}$ g cm^{-3} as the density at the level of the photosphere,[11] there is 10 times the flux required for heating the T_{min} network element bright points potentially available in Alfven waves generated from photospheric granular buffeting. The problem still remains, however, to show a viable dissipation mechanism for the T_{min} region.

Cook and Ewing[4] showed that the individual network bright points of 1–2 arc sec diameter were consistent with individual flux tubes with true, unresolved photospheric field strengths of around 1000–1500 G and diameters of ~1/3 arc sec, using a simple scaling law for the diameter d of a flux tube based on conservation of flux and equipartition of magnetic pressure and gas pressure p, $(d_2/d_1) = (p_1/p_2)^{1/4}$. The VAL model C atmosphere[11] gives the ratio of gas pressures at $\tau = 1$ and the T_{min} as 88, and so $d(T_{min}) = 3.1 d(photosphere)$. The observed Kitt Peak typical field strength in network bright points of ~50 G was obtained for ~2 arc sec seeing, and it is actually magnetic flux which is directly measured. Higher spatial resolution magnetograms by the Lockheed SOUP instrument[9] have reached 0.5 arc sec resolution, and found that field strengths associated with Ni I bright points are of the order of 600 G. Spatial resolution below 1/3 arc sec may show even higher field strengths for elementary photospheric flux tubes. The T_{min} is the highest level where network and active region features still keep a resolvable bright point structure. At greater heights the decreasing gas pressure and expanding fields cause the pointlike network features to spread out to more continuous patches of emission.

Our comparison with the energy available in Alfven waves was suggested by the observed linear relationship of excess heating to magnetic flux, but we did not mean to rule out other mechanisms which could yield such a result.

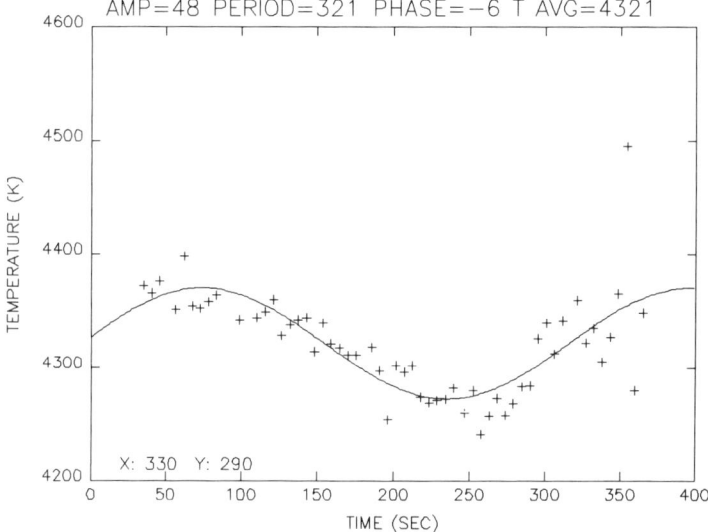

Fig. 4. Sine wave fit to brightness temperature variation for an individual 10 arc sec square box. The scatter to the fit illustrates the noise level of the data.

BRIGHTNESS OSCILLATIONS AT 1600Å

Cook and Ewing[5] also examined the brightness variations occuring in HRTS V 1600Å data in a series of 56 images over 330 s of the sounding rocket flight.[5] We found variations in small 10 arc sec patches in cell centers, which were immediately reminiscent of the 5 minute oscillation in the photosphere, and the oscillations observed by Noyes and Hall[7] in an IR line of CO at 4.67 μm which is formed in the T_{min}. I show in Figure 4 a sample result obtained by fitting the brightness temperature variations with time in individual 5 or 10 arc sec boxes filling a 350 x 350 arc sec field from our time series of 1600Å images. We used sine functions with individual average temperatures, amplitudes, periods, and phases for each box.

With only 330 s temporal coverage from our sounding rocket observations, a Fourier analysis ($\omega - \kappa$ diagram) was inappropriate. Instead we looked at the histogram distributions of the periods, amplitudes, etc., found for the individual

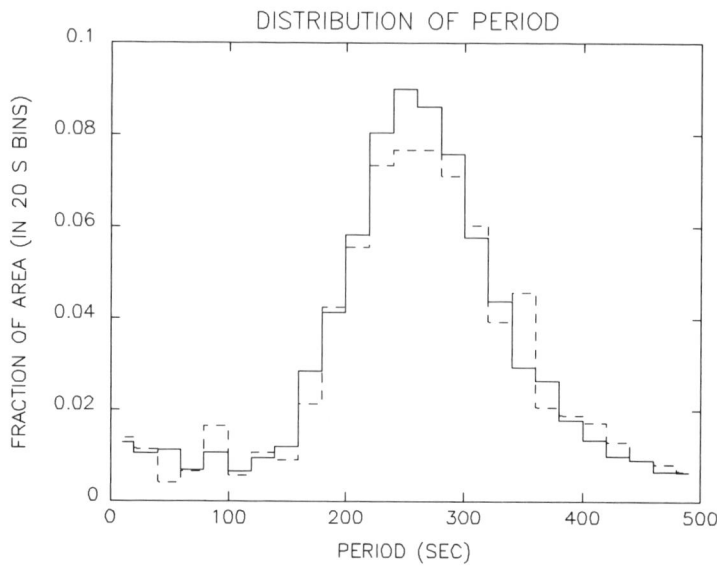

Fig. 5. Histogram distribution of period for 5 arc sec boxes (solid line) and for 10 arc sec boxes (dashed line).

5 or 10 arc sec boxes into which we divided our field. Figures 5 and 6 show the distribution of the periods and the amplitudes found from our sine fits to the variation of brightness temperature within individual boxes. The periods show a broad peak running from 150 to 400 s, centered at 250 s. A range of brightness temperature amplitudes was found, with the average of the 5 arc sec box distribution at 50 K.

We estimated the average energy flux $(1/2)\rho v^2 c_s$, where the sound speed $c_s = (\gamma kT/\mu m_H)^{1/2}$, for a simple propagating, adiabatic, undamped wave (however, it is not clear that these waves are actually propagating!). Other regions of the atmosphere depart from adiabaticity, but in the T_{min} the radiative relaxation time for radiative damping of compressional waves is usually estimated to be long compared with the compression time of an acoustic wave; in addition γ should be close to the monatomic ideal value of 5/3 because of the low degree of ionization. In this case $(v/c_s) = (\gamma - 1)^{-1}(\Delta T/T)$, and using $\Delta T = 50$ K gives an acoustic flux of 1.8×10^5 ergs cm^{-2} s^{-1}. This flux is small compared to an average global requirement of 8×10^6 ergs cm^{-2} s^{-1} to heat the T_{min} level,[1] and so the 250 s oscillations are energetically unimportant in the local heating budget.

To summarize these results, we suggested a picture for the quiet Sun in which a basal nonmagnetic heating and 250 s oscillations occur globally, while additional heating (perhaps from partial dissipation of the flux in Alfven waves produced from photospheric granular buffeting of individual flux tubes) produces the network bright points. These bright points may be individual elemental ~1500 G (at the photosphere) flux tubes which expand to arc sec diameter with decreasing gas pressure by the temperature minimum level. The average

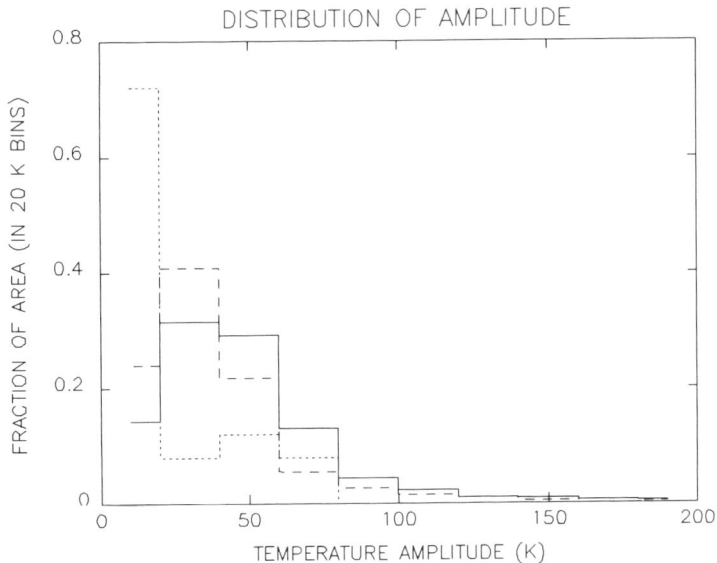

Fig. 6. Histogram distribution of amplitude for 5 arc sec boxes (solid line), 10 arc sec boxes (dashed line), and 70 arc sec boxes (fine dashed line).

additional energy flux required to maintain the network bright points is directly comparable to the flux required to heat the normal global average background at the temperature minimum level.

The relationship of brightness at 1600Å and magnetic flux in the photosphere must eventually flatten out above 50 G, as finally active region field strengths are reached. Already the brightest few elements in quiet areas are as bright as any features seen at 1600Å except for flares, and active regions simply contain a higher fractional area of these brighter elements. When field strengths characteristic of sunspots are reached, a presumably completely different physical mechanism takes over to actually suppress heating.

REFERENCES

1. L.S. Anderson and R.G. Athay, Ap. J. 346, 1010 (1990).
2. J.W. Cook, in Mechanisms of Chromospheric and Coronal Heating, ed. P. Ulmschneider, E.R. Priest, and R. Rosner (Berlin: Springer, 1991), p. 83-96.
3. J.W. Cook, G.E. Brueckner, J.-D.F. Bartoe, Ap. J. (Letters) 270, L89 (1983).
4. J.W. Cook and J.A. Ewing, Ap. J. 355, 719 (1990).
5. J.W. Cook and J.A. Ewing, Ap. J. 371, 804 (1991).
6. J.W. Cook, J.A. Ewing, and C.S. Sutton, Pub. A.S.P. 100, 402 (1988).
7. R.W. Noyes and D.N. Hall, Ap. J. (Letters) 176, L89 (1972).
8. N.R. Sheeley, Jr., J.D. Bohlin, G.E. Brueckner, J.D. Purcell, V. Scherrer, and R. Tousey, Solar Phys. 40, 103 (1975).

9. A. Title, T. Tarbell, K. Topka, D. Cauffman, C. Balke, and G. Scharmer, in The Physics of Magnetic Flux Ropes, ed. C.T. Russell (Washington: American Geophysical Union, 1990).
10. A.M. Title, T.D. Tarbell, K.P. Topka, S.H. Ferguson, R. A. Shine, and the SOUP Team, Ap. J. 336, 475 (1989).
11. J.E. Vernazza, E.H. Avrett, and R. Loeser, Ap. J. Suppl. 45, 635 (1981).

EXPLOSIVE EVENTS AND MAGNETIC RECONNECTION IN THE SOLAR ATMOSPHERE

K. P. Dere
E. O. Hulburt Center for Space Research, Code 4163
Naval Research Laboratory, Washington D. C. 20375, U.S.A.

ABSTRACT

Explosive events are highly-dynamic, small-scale phenomena commonly observed in spectra of transition zone lines. Their velocities are near 100 km s^{-1}, sizes near 1500 km, and time scales near 60 s. They occur at a height of 1000-2000 km, below typical transition zone structures. It has been demonstrated that some explosive events are caused by emerging magnetic flux and their is good evidence to indicate that the great majority are related to magnetic flux cancelation. The most probable mechanism for their generation is magnetic reconnection. The characteristics of explosive events and flux cancellation taken together show that magnetic reconnection in the quiet sun apparently proceeds readily in rapid bursts along the neutral line separating opposite flux elements that are convected together by photospheric flows.

INTRODUCTION

In 1975, the first rocket flight of the NRL High Resolution Telescope and Spectrograph took place. The ability of this instrument to simultaneously record high resolution spectra over a large field of view had an immediate impact on our view of the 'quiet' solar transition zone. These observations made it clear that the quiet transition zone was a highly dynamic environment. One of the more obvious examples of this sort of dynamic behavior were the newly discovered coronal jets and explosive events.[1] The coronal jets were characterized by velocities as high as 500 km s^{-1}, were roughly 10″ in size, and showed repeated accelerations at the same site. The explosive events were characterized by velocities of 100 km s^{-1} and spatial scales around 1500 km. They were much more numerous than the coronal jets. The deduced mass and energy associated with these events indicated that they potentially provided the source of the solar wind mass outflow and the heating for the quiet corona. Over the years, there have been 7 rocket flights of the HRTS and a Spacelab 2 mission in 1985. The analysis of these data sets have caused our ideas about coronal jets and explosive events to evolve. For one thing, it is now clear that the coronal jets are relatively rare events and were only fortuitously observed in the first two rocket flights. It seems unlikely that they occur sufficiently frequently to play a major role in the global balance of mass and energy in the solar corona. These later HRTS rocket flights have allowed a comparison of the locations and occurrences of the explosive events with photospheric magnetic fields and their evolution and this has brought about a much greater understanding of their basic nature. In this paper, I will then describe the basic properties of explosive events and the observations which point to their relationship with the reconnection of magnetic fields in the Sun.

PROPERTIES OF EXPLOSIVE EVENTS

Explosive events are observed as small-scale, impulsive features in the spectra of chromospheric, transition region, and coronal lines. They appear to be most prominent in transition region lines formed near 10^5 K where they most commonly appear in the HRTS spectra. Analyses of their properties as they appear in the HRTS spectra have been published previously.[1,2,3] In the HRTS spectra, the explosive events are detected as small features with large doppler shifts. Porter et al.[4] have reported small scale short period brightenings with large doppler signatures in SMM UVSP data that are probably closely related to explosive events seen in the HRTS spectra. Typical maximum velocities for the explosive events are 100 km s^{-1}. This can be compared with a sound speed of 50 km s^{-1} at 10^5 K. The profiles are usually asymmetric with the outflowing (blueshifted) and downflowing (redshifted) plasmas often displaced along the slit by 1000 km. Some of the profiles seem to be symmetric, perhaps indicating locations of high turbulence, but this is not usually the case. The redshifted profile is often different from the blueshifted profiles and in many examples, one of these components is often quite weak. On average, the velocities inferred from explosive event profiles show no dependence on their position with respect to either sun center or the limb. This implies that, in a statistical sense, the horizontal velocities (observed at the solar limb) and the vertical velocities (observed at disk center) are roughly equal. The velocities show no preferred direction and are in a sense isotropic. Times series of C IV spectra are available which show that over a time period of 200 s, no apparent motion of the explosive events along the slit is observed. If the explosive event were a single plasmoid moving at a velocity of 100 km s^{-1}, one would expect to detect motions of the spectral features along the slit. The observations place an upper limit of 5 km s^{-1} for the component of the plasmoid velocity across the solar surface.

The size of the explosive events tend to be around 1500 km. This is commensurate with the spatial displacements observed between blueshifted and redshifted components in the spectra. Structure on smaller spatial scales is also evident. It is clear that these are not just point-like explosions but three-dimensional objects whose structure is probably directly related to their dynamics.

In the HRTS spectra, the highest time resolution observations are found in a raster sequence with 20 s intervals between spectra at the same spatial location. This data set shows that the typical lifetime of an explosive event is 60 s but with strong changes often taking place with 20 s. Porter et al.[5] find transition brightenings in active regions with similar lifetimes and with some as short as 20 s. There is some evidence for repeatability of explosive events at the same site with some residual high nonthermal velocities between events. Explosive events observed in the quiet solar atmosphere are generally found at heights below typical transition region structures. This is demonstrated by the observations shown in Figure 1. These data were obtained on December 11, 1987 when two sounding rockets were launched, one carrying the AS&E X-ray telescope (D. Moses, P.I.) and the other carrying the NRL HRTS instrument for its fifth rocket flight (HRTS-5). Full disk magnetograms and He I λ 10830 spectroheliogram were also recorded by the Kitt Peak National Observatory. The data set obtained by this dual rocket flight is quite comprehensive, includes an image of the photospheric fields together with images spanning a full range of temperatures from the chromosphere, through the transition region and corona. The HRTS performed a raster sequence over a region of mostly quiet features, including the limb. The spectrograph was centered on the spectral region around 1550 which contains the C IV transition regions lines. Images of the intensity, velocity in the C IV lines were reconstructed from the spectra and coaligned with the full disk images. The locations of the explosive events found in the spectra were also placed onto a map for correlation with various images where they are indicated by small rectangles. Portions of these various images near the limb are shown in Figure 1.

Each image is $177'' \times 220''$. Directly at the limb, a number of explosive events are found at a height less than 2000 km just above the photosphere and below the limb brightening of the C IV transition regions lines. There has been general speculation that X-ray bright points, He I dark points, microflares and explosive events are related phenomena. This idea can be tested with the HRTS-5 data set. The locations of explosive events with respect to structures seen on the solar disk are shown in Figure 2. A number of explosive events seem to occur in the vicinity of the three X-ray bright points seen in this figure but the overall appearance of their distribution appears to be nearly random with regard to the X-ray bright points. Likewise, a number of explosive events occur inside He I dark points but there is no general correlation between the two phenomena. There is an indication that the explosive events are related to the supergranular network evident in the photospheric magnetogram, the He I spectroheliogram and the C IV intensity image.

In the quiet sun, a birthrate for the explosive events of 10^{-20} cm^{-2} s^{-1} has been derived from a sequence of 11 rasters obtained during the third HRTS rocket flight.[3] The birthrate in a polar coronal hole observed in the same flight was about half this number. A single large area raster obtained during the HRTS-7 rocket flight indicated that the explosive event birthrate in the quiet sun and in a coronal hole on the disk were equal.[8] One of the most important parameters for determining the relevance of explosive events on the global mass and energy balance of the solar coronal is their density. One method of determining the density is to calculate the emission measure ($\int n^2 dV$) and then divide by the volume of the feature. The main problem with this method is that it is possible for the volume of the emitting elements of the structure to be much smaller than its envelope. This is the general case for the quiet and active transition zone.[6] A more accurate measurement involves the use of density sensitive line intensity ratios, where the intensities of different lines from the same ion exhibit a different dependence on the density. This happens when metastable levels are present whose population does not vary linearly with density which is the general case with allowed lines directly excited from the ground level. The best line ratios in the HRTS spectral region for performing such an analysis for explosive events arise from the O IV ion formed near 1.7×10^5 K. Calculations show that line ratios involving these lines are sensitive indicators of electron density.[7] Unfortunately, they are not extremely intense lines and explosive events are not especially bright. One useable example where the O IV lines are bright enough to derive reliable densities has been found in the HRTS Spacelab 2 data. The line ratios in this event are consistent with a density of 7×10^{10} cm^{-3}. The average density obtained using the emission measure is

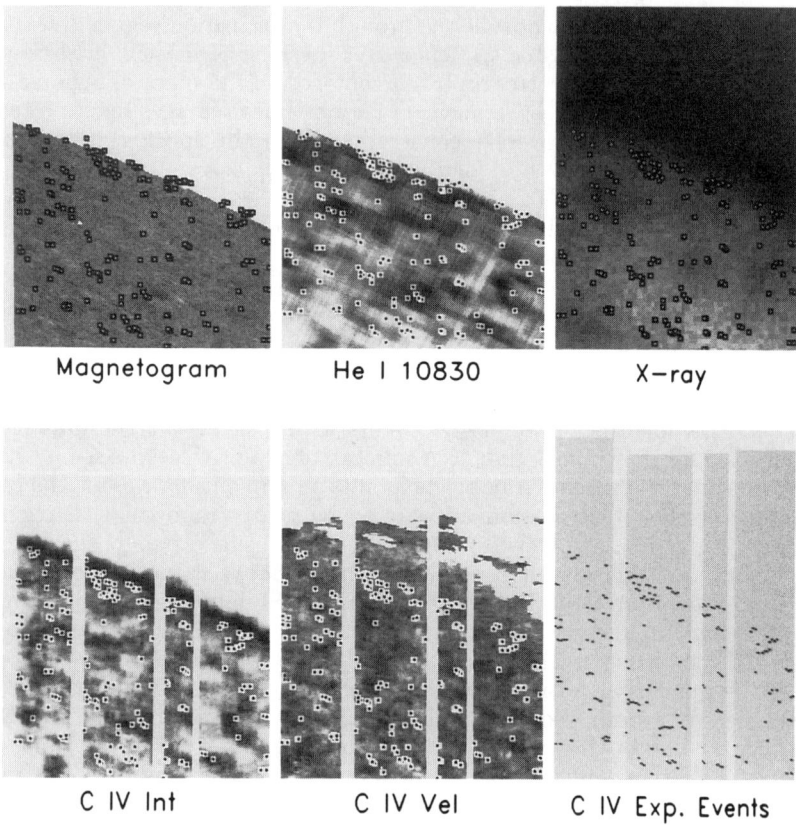

Fig. 1. Environment of explosive events near the solar limb.

1×10^9 cm^{-3}. The difference in the two density values indicates a volumetric fill factor of about 10^{-4}. The explosive event analyzed is an especially bright one but if the fill factors derived from this event apply to most explosive events then the masses and kinetic particle energies associated with explosive events much be reduced by this factor of 10^{-4} from the nominal values previously derived. As a consequence, the explosive events do not appear to play a direct role in the mass and energy balance of the corona. Nevertheless, they are probably a direct indicator of important processes occurring in the solar atmosphere that are directly connected to the mass and energy balance of the corona.

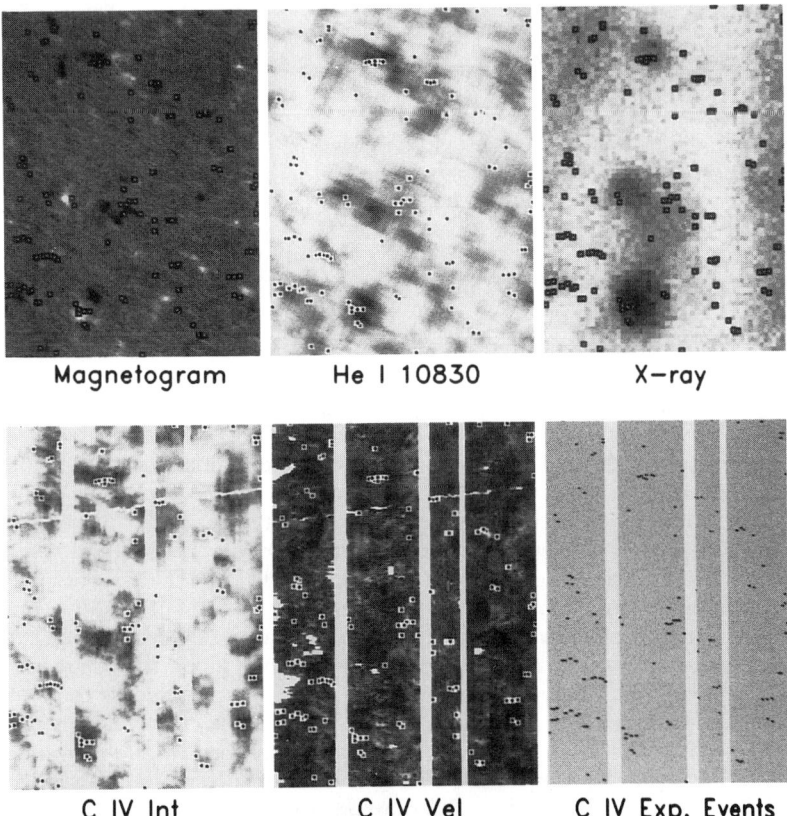

Fig. 2. Environment of explosive events near X-ray bright points on the solar disk.

Explosive events have typically been in observed in transition region lines formed at a temperature near 10^5 K. There are also a number of strong chromospheric lines in the HRTS spectra such as C I λ 1561 which should be sensitive to explosive event plasmas near 10^4 K. However, explosive events are only rarely seen in these lines. Although we have not attempted to determine the exact frequency, a rough estimate is that less that 1are also seen in chromospheric lines such as C I. They are seen weakly in C II λ 1335 formed at 4×10^4 K. Dynamic events very similar to the transition zone explosive events have been detected in coronal and upper transition zone lines observed with the NRL slitless spectroheliograph on Skylab.[9,10] Velocities are on the order of 200-500 km s^{-1}. Most of these events occur in active regions although some occur in quiet regions but their maximum temperatures are less than about 10^6 K. There are many fewer XUV streaks than transition region explosive events, even though the Skylab XUV spectroheliograph simultaneously observed the complete solar disk. One reason may be that only in the active regions is there enough plasma heating to generate the coronal temperatures.

THE ASSOCIATION OF EXPLOSIVE EVENTS WITH EMERGING AND CANCELING FLUX

It has generally been assumed that explosive events are driven by magnetic forces. For a variety of reasons, photospheric magnetic field data has been lacking until the last several rocket flights so that it has not been possible to delve into this question. The first clues came from HRTS C IV spectra over an active region obtained during Spacelab 2. These showed unusually high velocities and brightness over an extended region that was associated with an emerging flux region seen in Hα spectroheliograms obtained at the Hida observatory.[11] The association of explosive events with emerging flux was further demonstrated with data obtained during the sixth HRTS rocket flight (HRTS-6). The primary target for this flight was a coronal hole on the disk but the slit raster sequence also included a small emerging active region. This active region was a prolific producer of explosive events, most of which were aligned along the boundaries of strong flux regions with many on the neutral line.[8] These data led us to conclude that the explosive events were produced by the emergence of new magnetic flux into the solar atmosphere. In particular, this would proceed by the reconnection of magnetic flux as in the Heyvaerts, Priest and Rust flare model.[12] The choice of reconnection is supported by the bipolar nature of the reconnection process which would drive material out of the reconnection region in opposite directions as observed in the explosive events.

It does not appear to be possible to associate most of the explosive events with magnetic flux emergence. Under the assumption that the explosive events are associated with magnetic reconnection, the best candidate for the production of the explosive events are the magnetic flux cancellation events. Flux cancellation was discovered by Martin[13] and is described by Martin et al.[14] and Livi et al.[15] The basic picture is that small magnetic bipoles emerge in the supergranular cell centers and are transported to the boundaries where opposite polarity flux elements are driven together by the photospheric flows and forced to cancel.

The identification of flux cancellation events as the source of the explosive events in the quiet sun is supported by the fact that many explosive events occur on the edges of the supergranular network defined both by the C IV line intensities and by the photospheric fields. This can be seen in the HRTS-5 data seen above and also in the HRTS-6 data set.[8] At first glance, the explosive events appear to be randomly distributed over the quiet sun. After some study, their distribution can be described by several general characteristics. They tend to avoid the regions of highest photospheric magnetic fields strength. They frequently occur on the borders of the high field regions, often just outside the high field area where the observed field strength appears to be negligible. Many other are just scattered about the solar surface in places where there appears to be no observable field, probably because the combined resolution and sensitivity of the available observations during the HRTS flights have not been able to detect the weaker bipolar fields that are generally present in the absence of strong fields.

DIAGNOSTICS OF MAGNETIC RECONNECTION

The characteristics of magnetic cancellation and explosive events can be used to shed some light on the nature of magnetic reconnection as it occurs in the quiet sun outside of flares. For example, if the velocity observed in an explosive event is equated with the local Alfévn speed, then a knowledge of the plasma density leads to a value for the magnetic field strength in the reconnection region. For the single event where the electron density has been measured, a magnetic field strength of 20 gauss is derived. This value seems to be roughly comparable with photospheric field strengths measured with BBSO magnetograms of canceling intranetwork flux elements.[15]

The cancellation of photospheric flux is observed to occur across boundaries on the order of 6000 km and on time-scales of several hours. A simple model suggests that cancellation then occurs as one flux element is driven into another at a velocity around 0.3 km s^{-1}, very close to the velocity observed for the convection of flux elements through the supergranular cell interior.[16] The explosive events are characterized by size scales of 1500 km and time scales of 60 s, much smaller and shorter than observed for the process of flux cancellation. This suggests that the actual reconnection occurs in short bursts at discrete locations along the overall interaction boundary. This is graphically summarized in Figure 5. The rate at which reconnection proceeds is quite close to the rate predicted by the Petschek mechanism.[17]

The net effect of these flux cancellation/reconnection events on the overall structure and energetics of the solar atmosphere is not immediately clear. With regard to mass transport in the corona, the spectra certainly show that jets directed outward from the solar surface are produced. Some of the jets will occur on field lines that are newly opened to the corona by the reconnection event. The material in these jets can escape into the solar wind. Because the densities are high (at least in one case), the total mass associated with an explosive event at 10^5 K is not high enough for them to play a strong role in the supply for the solar wind. One possibility is that the mass deduced from C IV observations can account for the total mass in the event. Plasma heating does result from the reconnection that accelerates the is involved since 10^5 K material is formed at a height where much cooler material previously existed. The observations reported by Cheng and Kjeldseth-Moe[10] suggest that in most events, the material is not heated to coronal temperatures. The energy losses in the quiet sun are dominated by the high density structures in the strong network field regions but the explosive events mainly occur on the boundaries of these strong field regions. Parker[18] suggests that magnetic reconnection in coronal holes may be responsible for the heating in these regions and for the generation of Alfévn waves which accelerate the high speed wind streams. Dere[19] has deduced the existence of considerable power at wavelengths short enough to be dissipated. It is possible that this power is the result of reconnection processes occurring over a range of spatial scales. Another result of the network boundary reconnection events would be to enhance the rate at which magnetic complexity is built up by the shuffling of the magnetic field lines that Parker[20] has suggested as the root source of free energy in the corona.

REFERENCES

1. G. E. Brueckner and J.-D. F. Bartoe, Astrophys J. 272, 329 (1983).
2. J. W. Cook, P. A. Lund, J.-D. F. Bartoe, G. E. Brueckner, K. P. Dere, and D. G. Socker, Cool Stars, Stellar System, and the Sun, eds. J. L. Linsky and R. E. Stencel, Lecture Notes in Physics 291 (Springer-Verlag, 1987), p. 150.
3. K. P. Dere, J.-D. F. Bartoe, G. E. Brueckner, Solar Phys. 123, 41 (1989).
4. J. G. Porter, R. L. Moore, E. J. Reichmann, O. Engvold, and K. L. Harvey, Astrophys. J. 323, 380 (1987).
5. J. G. Porter, U. Toomre, and K. B. Gebbie, Astrophys. J. 283, 879 (1984).
6. K. P. Dere, J.-D. F. Bartoe, G. E. Brueckner, J. W. Cook, and D. G. Socker, Solar Phys. 114, 223 (1987).
7. K. P. Dere, J.-D. F. Bartoe, G. E. Brueckner, Astrophys. J. 259, 366 (1982).
8. K. P. Dere, J.-D. F. Bartoe, G. E. Brueckner, J. Ewing, and P. Lund, J.G.R. 96(A6), 9399 (1991).
9. G. E. Brueckner, N. P. Patterson, and V. E. Scherrer, Solar Phys. 47, 127 (1976).
10. C.-C. Cheng and O. Kjeldseth-Moe, Dynamics of Solar Flares, eds. B. Schmieder and E. Priest (Observatoire de Paris, DASOP, 1991), p. 101.
11. G. E. Brueckner, J.-D. F. Bartoe, J. W. Cook, K. P. Dere, D. G. Socker, H. Kurokawa, M. McCabe, Astrophys J. 335, 986 (1988).
12. J. Heyvaerts, E. R. Priest, and D. M. Rust, Astrophys. J. 216, 123 (1977).
13. S. F. Martin, Small-Scale Dynamical Processes in Quiet Stellar Atmosphere, ed. S. L. Keil (National Solar Observatory, 1984), p. 30.
14. S. F. Martin, S. H. B Livi, and J. Wang, Aust. J. Phys. 38, 929 (1985).
15. S. H. B. Livi, J. Wang, and S. F. Martin, Aust. J. Phys. 38, 855 (1985).
16. H. Zirin, Aust. J. Phys. 38, 961 (1985).
17. H. E. Petschek, The Physics of Solar flares, ed. W. N. Hess (NASA SP-50, 1964), p. 425.
18. E. N. Parker, Astrophys. J. 372, 719 (1991).
19. K. P. Dere, Astrophys. J. 340, 599 (1989).
20. E. N. Parker, Astrophys. J. 330, 474 (1988).

THREE-DIMENSIONAL KINEMATIC RECONNECTION OF PLASMOIDS WITH NULLS

Yun-Tung Lau
Code 930.1, NASA, Goddard Space Flight Center, Greenbelt, MD 20771, USA.

John M. Finn
Lab. for Plasma Research, Univ. of Maryland, College Park, MD 20742, USA.

ABSTRACT

The global nonlinear dynamics of magnetic field lines in plasmoids with a pair of nulls, where $\boldsymbol{B} = 0$, is studied. The aim of this analysis is to describe the separatrix surfaces on which singularities can occur in ideal magnetohydrodynamics because of topological changes in the field. These separatrix surfaces should locate the boundary layers associated with three-dimensional reconnection in the presence of resistivity or inertia. It is found that the field lines exhibit chaotic scattering with several properties in common with plasmoid models without nulls (in which one component of the magnetic field never changes sign). In particular, the singular surfaces can be fractal, implying complex current density structures down to the dissipation scale. These generic features are expected to exist in typical coronal magnetic geometries exhibiting three-dimensional reconnection and the formation of current sheets.

INTRODUCTION

In reference 1 we suggested an approach called kinematic reconnection to investigate the qualitative features of magnetic reconnection in three dimensions. In this type of analysis, one takes a magnetic field $\boldsymbol{B}(\boldsymbol{x}, t)$ with an associated vector potential $\boldsymbol{A}(\boldsymbol{x}, t)$, having $\nabla \times \boldsymbol{A} = \boldsymbol{B}$. Then one computes the scalar potential ϕ such that the electric field $\boldsymbol{E} = -\partial \boldsymbol{A}/\partial t - \nabla \phi$ can satisfy the ideal magnetohydrodynamic (MHD) condition

$$\boldsymbol{E} + \boldsymbol{v} \times \boldsymbol{B} = 0, \tag{1}$$

namely by solving

$$\boldsymbol{B} \cdot \nabla \phi = -\boldsymbol{B} \cdot \frac{\partial \boldsymbol{A}}{\partial t}.$$

If ϕ is nonsingular, the fields can in principle evolve via ideal MHD. On the other hand, singularities in ϕ occur in areas where the specification of $\boldsymbol{B}$ and of the inductive electric field

$$\boldsymbol{E}_{\text{in}} = -\frac{\partial \boldsymbol{A}}{\partial t}$$

indicate topological changes in the field lines, which are not allowed in ideal MHD. In the presence of resistivity or inertia, these singularities should locate the boundary layers, or current sheets, in reconnection.[1] A simple example is a two-dimensional X-point, in which singularities occur along the X-shaped separatrix. More generally, singularities are associated with field nulls,[1,2] closed field lines,[1] or field lines that form nested tori.[3]

The local properties of the singularities with closed field lines[1] or nulls[1,2] has been studied. More recently, we investigated the global properties of plasmoids with infinitely long (or closed) field lines[3] These plasmoid models do not allow nulls, because one component of $\boldsymbol{B}$ never vanishes. In that case, we found that the singularities in ϕ are associated with the nonlinear dynamical structures of the field line equations, which form a Hamiltonian system, and that the results can be described in terms of Hamiltonian chaotic scattering.

In this paper we use the time delay functions to explore the structure of singularities in plasmoid models with nulls. Details of the nonlinear dynamics of the field line equations, especially the bifurcations, which change the topology of the field lines, are analysized in a separate paper[4].

PLASMOID MODELS AND TIME DELAY FUNCTIONS

In reference 3 we introduced three models for plasmoid-like fields, appropriate to solar coronal structures or to the magnetotail. These models are the short plasmoid, the long plasmoid, and the periodic plasmoid. They represent a straight arcade aligned along z where plasmoid-like perturbations have occurred locally along its length, all along its length, or periodically along its length, respectively. Their magnetic fields are given by

$$\boldsymbol{B} = \boldsymbol{\nabla}\psi(x,y,z) \times \hat{\boldsymbol{e}}_z + B_z(x,y)\hat{\boldsymbol{e}}_z, \qquad (2)$$

with $B_z = B_0 =$ constant. There is no qualitative difference if $B_z(x,y)$ is nonuniform but does not change sign.[3] (If these are considered as models for the magnetotail fields, B_z is the dawn-dusk component, usually called B_y.) The short plasmoid has ψ surfaces (not magnetic surfaces, since $\boldsymbol{B} \cdot \boldsymbol{\nabla}\psi \neq 0$) as shown in figures 1a and 1b.

For $|z|$ larger than the plasmoid length, ψ is arcade-like. A simple model has $A_z = \psi - E_0 t$ and

$$\psi = y^2/2 + (x-1)^3/3 + (z^2 - 1/4)x,$$

which gives

$$B_x = y, \quad B_y = -(x-1)^2 - z^2 + \frac{1}{4}, \qquad (3)$$

and $\boldsymbol{E}_{\text{in}} = E_0 \hat{\boldsymbol{e}}_z$. The x-z plane with the circle where $B_y = 0$ (which surrounds a finite region for a short plasmoid) is shown in figure 1c. In the short plasmoid, there are no singularities in ϕ, which is proportional to the distance a field line travels in z for E_0 constant. (The singularities in ϕ lie on a complex surface for the long plasmoid, and on a spatially intermittent, fractal structure for the periodic plasmoid.[3]) This property of the short plasmoid is traced to the fact that all field lines are topologically equivalent, which is because there are no nulls, closed field lines, or nested tori.

However, there are plausible conditions under which B_z might change sign. Indeed, the dawn-dusk component of the field in the magnetotail has been observed to have sign changes.[5] Also, certain models for solar arcade-like structures, which develop plasmoids because of an MHD instability, have B_z that

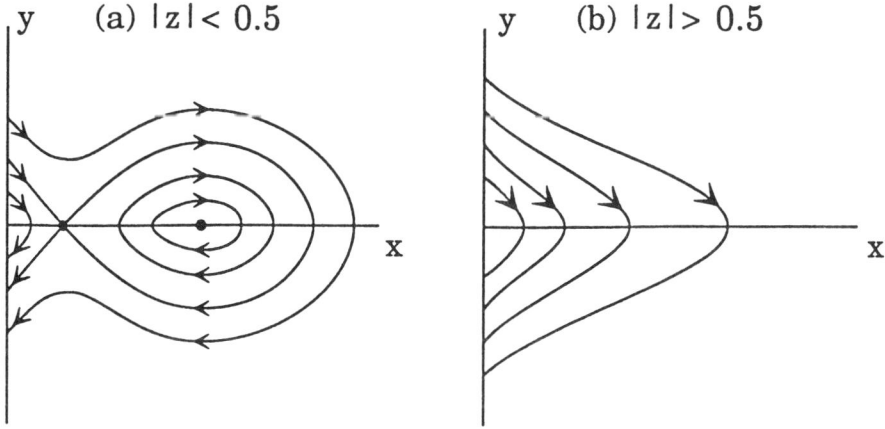

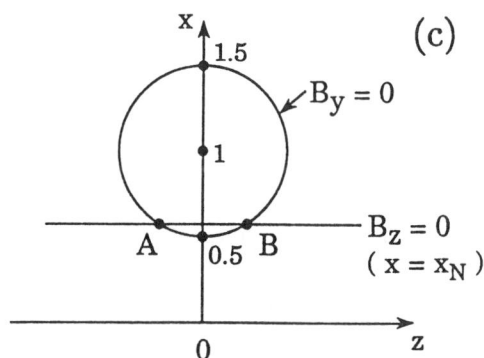

Fig. 1. Surfaces of constant ψ in the x-y plane for equation (3) for (a) $|z| < 0.5$, (b) $|z| > 0.5$; (c) the x-z plane for $y = 0$ showing the circle where $B_y = 0$ and the line where $B_z = 0$.

changes sign.[6] If this occurs, nulls as well as closed field lines can exist, leading to singularities. As a model, we take

$$B_z = B_0(x - x_N). \tag{4}$$

The field described by equations (3) and (4) has nulls at $x = x_N$, $y = 0$, and $z = \pm z_N = \pm[(1/4 - (x_N - 1)^2]^{1/2}$ if $0.5 < x_N < 1.5$, as indicated in figure 1c. The linearized field line equations near a null take the form $d\delta r/d\tau = \delta r \cdot \nabla B$, where δr is measured from the null, ∇B is a 3×3 matrix evaluated at the null,

74 Plasmoids with Nulls

and τ is an artificial time-like variable increasing along field lines. This matrix has three distinct real eigenvalues if $8(1-x_N)^3 > 27B_0^2[1/4-(1-x_N)^2]$. If the reverse is true, then it has one real and two complex conjugate eigenvalues, and the null is called a spiralling null. Furthermore, the null at $z = -z_N$ is of type A because the signs of the eigenvalues (the real parts of the eigenvalues in the spiralling case) are $(+ - -)$.[1,2,7] On the other hand, the null at $z = +z_N$ is of type B since the signs are $(- + +)$.

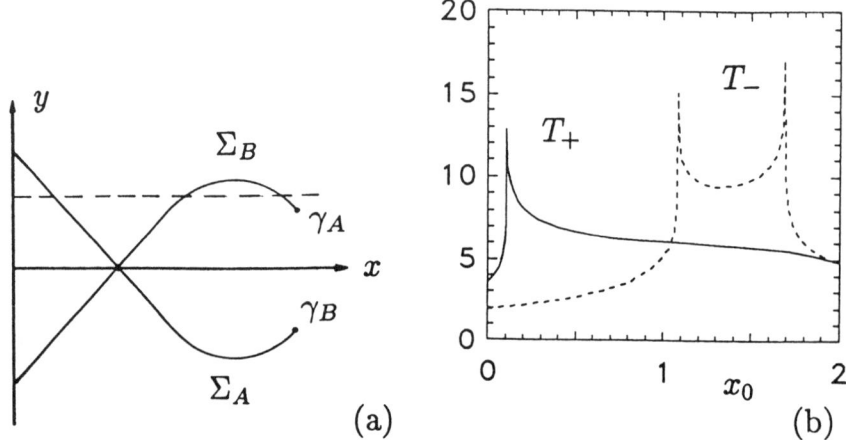

Fig. 2.   Results for $B_0 = 0.19245$ and $x_N = 0.6$ (nonspiralling nulls). (a) A sketch of the intersection of Σ_A, γ_A, Σ_B, and γ_B with the x-y plane. (b) Time delay functions T_- (dotted) and T_+ (solid) versus x_0 for $y_0 = 0.45$ [the dashed line in (a)].

For a type A null, there are two field lines (together called γ_A) emanating from the null. Near the A null, they are parallel to the eigenvector with positive eigenvalue. Starting from a point in space, one can follow a field line by solving the field line equations $d\boldsymbol{r}/d\tau = \boldsymbol{B}(\boldsymbol{r})$. When the field line passes near A, it typically leaves A along either arm of γ_A. If it does not, then it must approach A asymptotically as $\tau \to \infty$. The collection of such starting points that land on A forms a two-dimensional surface in space, called Σ_A. Σ_A is like a separatrix surface that separates field lines approaching either arm of γ_A. Similarly, there are two field lines γ_B approaching the B null, connected to the eigenvector with negative eigenvalue. And if we go backward in (fictitious) time, there is a separatrix surface Σ_B that separates field lines going along either arm of γ_B.

The location of the separatrix surfaces can be detected with the time delay functions

$$T_\pm(x_0, y_0, z_0) = \int_{\boldsymbol{r}_0}^{|\boldsymbol{r}_1|=R} d\tau.$$

Here T_+ is for integration forward in time and T_- for backward. Typically we take a boundary sphere with radius $R = 20$. In the present paper, we restrict starting points to the x-y plane, i.e., $z_0 = 0$. When integrated forward, points (x_0, y_0) near Σ_A (γ_B) come very near the A null (the B null) and take a long time $T_+(x_0, y_0)$ to reach the boundary sphere, giving a logarithmic singularity at the intersection of the x-y plane with Σ_A (γ_B). Similarly, the time $T_-(x_0, y_0)$ for integrating backward has logarithmic singularities on Σ_B and γ_A. The potential ϕ of kinematic reconnection has the formal solution

$$\phi = \int d\tau\, \boldsymbol{B} \cdot \boldsymbol{E}_{\text{in}}$$

and therefore its singularities are related to those of $T_\pm$. Consequently, computing $T_\pm$ and identifying its singularities yields the separatrix surfaces, which are associated with boundary layers in reconnection.

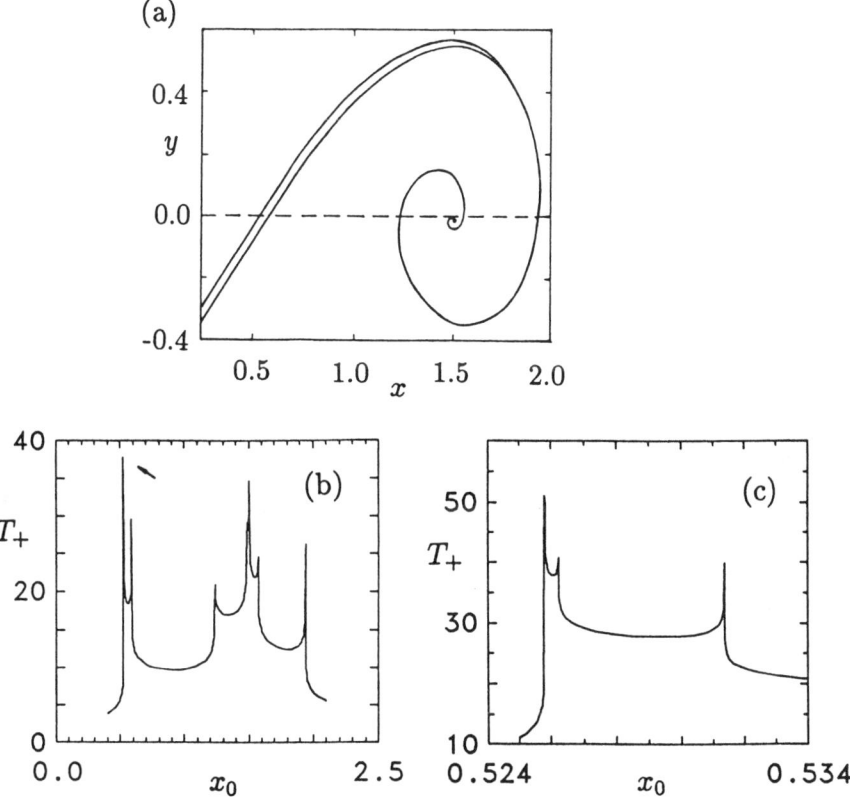

Fig. 3. Results for $x_N = 1.1$ (spiralling nulls). (a) Intersection of the singularities of T_- (Σ_B) with the x-y plane (T_+ gives a mirror image with respect to the x-axis). (b) The time delay function T_+ versus x_0 for $y_0 = 0$. (c) A blowup of the leftmost spike in (b) (indicated by an arrow), showing the fractal nature of the singularities.

NUMERICAL RESULTS

For the field given by equations (3) and (4) with $B_0 = 0.19245 \approx 1/\sqrt{27}$ (this value of B_0 will be used throughout the paper), we numerically compute the time delay function for increasing values of x_N between 0.5 and 1.5. In figure 2a we show a sketch of the intersection of Σ_A, γ_A, Σ_B, γ_B with the x-y plane. Note that γ_A is on the boundary of Σ_B and γ_B on the boundary of Σ_A.[2] Because our model is symmetric under $y \to -y$, $z \to -z$, $\tau \to -\tau$, Σ_B (γ_A) is the reflection of Σ_A (γ_B). Furthermore, Σ_A and Σ_B intersect the x-axis at the same point in figure 2a $[T_-(x_0, 0) = T_+(x_0, 0)]$. Such intersections are the B-A lines, which corresponds to a field line that goes from B to A. There is only one B-A line for $x_N = 0.6$. In figure 2b, we show $T_\pm$ for a cut of initial conditions $0 < x_0 < 2$ and $y_0 = 0.45$. The logarithmic singularities at the separatrix surfaces are evident.

As x_N is increased, the separatrix surfaces become increasingly complicated. In figure 3a, we show the locus of the singularities of T_- (Σ_B) on the x-y plane for $x_N = 1.1$ (Σ_A is the mirror image of Σ_B with respect to the x-axis). For these parameters, there is more than one intersection of Σ_A with the x-axis and therefore more than one B-A line, and an apparent doubling of the Σ surfaces. Figure 3b shows a cut for $0.0 < x_0 < 2.5$ and $y_0 = 0$. Figure 3c is an expansion of the leftmost spike in figure 3b, showing a fractal distribution of singularities. The corresponding Σ surfaces are fractal (with an infinite number of B-A lines) and the field lines are chaotic. This change in topology is realized through a series of global bifurcations.[4,8] For example, an unstable closed field line is created when γ_A and γ_B first meet on the x-axis. In that case, the chaos is associated with the nulls and the unstable closed field line. The fact that $T_\pm$ is singular on a fractal set indicates that the field lines exhibit chaotic scattering,[9,10] in which final quantities related to field lines leaving the system are extremely sensitive to the initial conditions.

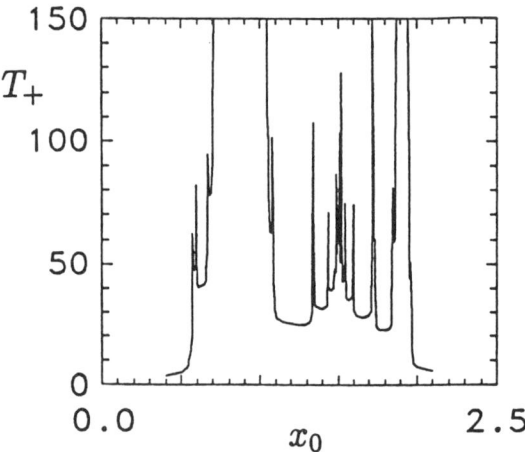

Fig. 4. T_+ for $x_N = 1.32$ ($y_0 = 0$), showing the "castle"-like fractal.

In figure 4 we show $T_\pm$ for $x_N = 1.32$. For this set of parameters there is a "castle" region, as observed in the periodic plasmoid.[3,11] This castle structure indicates the existence of invariant KAM tori (quasiperiodic orbits) and represents the field lines that explore the "sticky" region near the last KAM tori. There is a stable (elliptic) field line at the center of the tori, and the invariant tori are formed when the unstable closed orbit becomes stable by an inverse period doubling bifurcation. In figures 5a and 5b we show surfaces of section, i.e., a plots of successive crossings of the $y = 0$ plane and the $z = 0$ plane, respectively. Chaotic scattering persists in the area outside the KAM tori. This chaotic scattering is due to the multiple B-A lines, and has nothing to do with the closed field line at the center. Within the region of KAM tori, one can find a nonsingular change of variables under which the qualitative behavior is equivalent to that in a Hamiltonian system.[3,4]

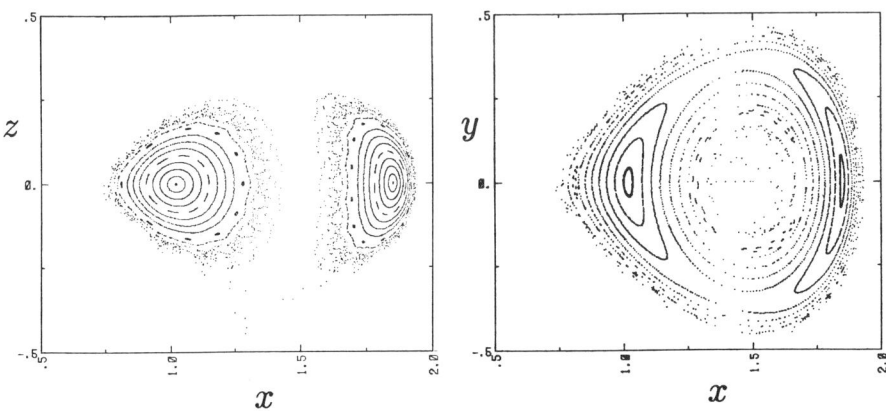

Fig. 5. Surfaces of section for $x_N = 1.4$. (a) for $y = 0$ and (b) $z = 0$.

As $x_N \to 1.5$ the nulls approach each other and the whole structure self similarly shrinks, with dimensions scaled as $(1.5 - x_N)^{1/2}$. For $x_N > 1.5$ there are no nulls or closed field lines and all orbits go to infinity after a finite number of crossings of the $z = 0$ plane. That is, separatrix surfaces no longer exist.

In summary, we have found that the global nonlinear dynamics of plasmoids with nulls in three dimensions is different from, but shares many features with, periodic plasmoids in which a component of B never vanishes (Hamiltonian case). Chaotic scattering is observed by means of the time delay function and has several properties in common with periodic plasmoids. In particular, for a range of parameters, the fractal nature of the separatrix surfaces indicates a complex web of boundary layers, or current sheets, for reconnection. The fractals are created by global bifurcations. The field line chaos can be associated with the nulls alone, or with the nulls in conjunction with closed field lines.

ACKNOWLEDGMENTS

The work of Y.T.L. was done while the he held a National Research Council-NASA/GSFC Research Associateship. The work of J.M.F. was supported by NASA grants NAGW-1346 and NAG5-1356 and by the U.S. Department of Energy.

REFERENCES

1. Y.-T. Lau and J. M. Finn, Ap. J. 350, 672 (1990).
2. J. M. Greene, J. Geophys. Res. 93, 8583 (1988).
3. Y.-T. Lau and J. M. Finn, Ap. J. 366, 577 (1991).
4. Y.-T. Lau and J. M. Finn, submitted to Physica D , (1991).
5. W. J. Hughes and D. G. Sibeck, Geophys. Res. Lett. 14, 636 (1987).
6. J. M. Finn, P.N. Guzdar, and J. Chen, submitted to Ap. J. , (1992).
7. S. W. H. Cowley, Radio Science 8, 903 (1973).
8. J. Guckenheimer and P. Holmes, Nonlinear Oscillations, Dynamical Systems and Bifurcation of Vector Fields (Springer-Verlag, Berlin, 1983).
9. B. Eckhardt, Physica D 33, 89 (1988).
10. U. Smilansky, in Lectures at Les Houches, Chaos and Quantum Physics, edited by M. -J. Giannoni, A. Voros and J. Zinn-Justin (Elsevier, Amsterdam, 1990).
11. Y.-T. Lau, J. M. Finn, and E. Ott, Phys. Rev. Lett. 66, 978 (1991).

GENERATION OF MAGNETIC FIELDS BY CHAOTIC FLUID CONVECTION: THE FAST DYNAMO PROBLEM

John M. Finn
Lab. for Plasma Research, Univ. of Maryland, College Park, MD 20742

There has been active investigation on the fast dynamo problem in the last few years, initiated by the seminal observation made by Arnold et al.[1] that the dynamo growth rate should be related to the positive Lyapunov exponent in chaotic flows, i.e., in flows with chaotic streamlines.

Recent work on the fast dynamo has split into two categories: numerical simulations of the equations of motion in "realistic" flows such as the ABC flow[2,3] and approaches using more idealized flows and relating to the underlying nonlinear dynamics of the flow.[4-10] I will not attempt a systematic review of these developments here. Rather, I will present an informal review, preferring to linger on some topics of a more intuitive or speculative nature and not shying away from personal opinions. And now we must begin with some definitions.

A dynamo is a mechanism for creating magnetic fields by the stretching involved in a flow in a conducting fluid or plasma. The kinematic approach to the study of dynamo theory takes the flow $v(x,t)$ to be given, as is appropriate if the magnetic field is (initially) very small. In this limit, the Lorentz force $j \times B$ in the equation of motion is negligible, and the flow is not influenced by the field. In this kinematic limit, the magnetic field obeys a linear equation [Eq. (1) below] and we "have a dynamo" if the field has a positive growth rate. As we shall discuss below, a chaotic flow is one with chaotic streamlines, i.e., one with a positive Lyapunov exponent. That is, the chaos is in the Lagrangian orbits of the flow $v(x,t)$ and not in its explicit eulerian dependence. Finally, a fast dynamo is a kinematic dynamo whose growth rate remains finite in the limit of zero electrical resistivity η. A dynamo whose growth rate γ approaches zero as $\eta \to 0$ is called a slow dynamo.

Because of its sheer size, and because it is highly conducting, the solar convection zone has an extremely large magnetic Reynolds number $R_m \gtrsim 10^{10}$. ($R_m = vL/\eta$ is the magnetic Reynolds number, where v and L are nominal velocity and length scales, respectively.) Because of this, it is thought that only fast dynamos are relevant, although it is conceivable that anomalous resistivity or collisionless effects, as in collisionless reconnection, may play a role.

In the kinematic dynamo problem, the velocity $v(x,t)$ is prescribed, and the basic equation of motion for the magnetic field follows from Ohm's law and Faraday's law:

$$\frac{\partial B}{\partial t} = \nabla \times (v \times B) - \nabla \times (\eta \nabla \times B),$$
$$= -v \cdot \nabla B + B \cdot \nabla v + \eta \nabla^2 B, \qquad (1)$$

where η (taken to be constant) is the dimensionless resistivity ($\eta = R_m^{-1}$). We have assumed incompressible flows $\nabla \cdot \boldsymbol{v} = 0$. Since the equation

$$\frac{\partial \psi}{\partial t} + \boldsymbol{v} \cdot \nabla \psi = \eta \nabla^2 \psi \tag{2}$$

is called the "passive scalar" problem, the dynamo equation (1) can be called the "passive vector" problem.

The equation of motion for a fluid element in the plasma is

$$\frac{d\boldsymbol{x}_0(t)}{dt} = \boldsymbol{v}(\boldsymbol{x}_0(t), t) \tag{3}$$

and its variational equation (i.e., equation for the difference $\delta\boldsymbol{x} = \boldsymbol{x}_1(t) - \boldsymbol{x}_0(t)$ between two nearby orbits) is

$$\frac{d}{dt}\delta\boldsymbol{x}(t) = \delta\boldsymbol{x}(t) \cdot \nabla \boldsymbol{v}(\boldsymbol{x}_0(t), t). \tag{4}$$

Note that Eq. (1) for $\boldsymbol{B}$ with $\eta = 0$ is identical to Eq. (4) with $\delta\boldsymbol{x} = \boldsymbol{B}$. The magnetic field is convected along by the flow and stretched by the action of the term $\boldsymbol{B} \cdot \nabla \boldsymbol{v}$, the term on the right in (4).

In the slow dynamo of Childress[11] and of Perkins and Zweibel,[12] the flow is of the form

$$\boldsymbol{v} = \nabla\phi \times \hat{\boldsymbol{e}}_z + \lambda\phi\hat{\boldsymbol{e}}_z, \tag{5}$$

where $\phi = \sin x \sin y$. Here $\boldsymbol{v}$ is a Beltrami flow, satisfying $\nabla \times \boldsymbol{v} = \lambda \boldsymbol{v}$ with $\nabla^2 \phi + \lambda^2 \phi = 0$ with $\lambda^2 = 2$. This flow is two dimensional, and therefore the field satisfying (1) is of the form $\delta\boldsymbol{B}(x,y)\exp(ikz)$. There can be no dynamo for $k = 0$ by Cowling's theorem. For $k \neq 0$ there is a dynamo, and the growth of the magnetic field is associated with the strain near the stagnation points at $(x, y) = (m\pi, n\pi)$. In fact $\delta\boldsymbol{B}$ must be of the form

$$\delta\boldsymbol{B}_p(x,y)e^{i\boldsymbol{k}\cdot\boldsymbol{x}},$$

where $\delta\boldsymbol{B}_p$ is periodic in x, y of period 2π. The "α-effect" calculation of Refs. 11 and 12 assumes a separation of scales $|\boldsymbol{k}| \ll 1$, but this condition occurs for computational convenience and not for any fundamental reason. Because of the integrable nature of the flow (5), the field[11,12] is confined to a boundary layer near the separatrices $\{x = m\pi\}$ and $\{y = n\pi\}$ of width of order $\eta^{1/2}$, giving a dynamo growth rate $\gamma \sim \eta^{1/2}$. Thus, this is not a fast dynamo. There is no fundamental importance of stagnation points. For example, $\phi \to \phi + C$, where C is a constant, produces a flow without stagnation points but with x-lines at $\{x = m\pi, y = n\pi\}$. (From this translating reference frame $\Delta v_z = \lambda C$, the complex frequency is simply doppler shifted.) Rather, what is important is the exponential stretching near the x-lines.

A chaotic flow is one in which there is sensitive dependence to initial conditions in (3); that is, solutions $\delta\boldsymbol{x}(t)$ of the variational equation (4) have a positive Lyapunov exponent

$$\lambda_L \equiv \lim_{t\to\infty} \frac{1}{t}\ln\left[\frac{|\delta\boldsymbol{x}(t)|}{|\delta\boldsymbol{x}(0)|}\right] > 0. \tag{6}$$

A two dimensional steady flow [of which (5) is an example] cannot be chaotic, but time dependence or dependence on a third spatial variable allows chaos to form in an "ergodic" region of space. (There can be distinct ergodic components, divided by invariant surfaces, and having distinct Lyapunov exponents.)

Fast dynamos have been observed in chaotic flows, and their essence seems to be the following: in a chaotic flow, there is no separatrix, as there is for the flow of Eq. (5). Rather, the stable and unstable manifolds of stagnation points, or of closed unstable streamlines, form tangles filling the ergodic regions. Then the $\eta^{1/2}$ boundary layers overlap in the ergodic regions. Thus, the field can occupy a larger region than the individual boundary layers of the slow dynamo, and its growth rate can become independent of η as η approaches zero. As observed in Ref. 7, the length scales decrease exponentially during the early (ideal MHD) phase, until they reach a scale $|\Delta \boldsymbol{x}| \sim R_m^{-1/2} = \eta^{1/2}$, after which an eigenfunction sets up. For $R_m \gg 1$, this length scale is very much smaller than the length scale of any macroscopic area.

As in the case with integrable (nonchaotic) streamlines, the dynamo here can be due to chaos associated with unstable stagnation points or unstable closed (periodic) orbits. From a translating or rotating reference frame, a stagnation point again becomes a periodic orbit, but in this case a steady flow develops a time dependence. However, in Ref. 8 it has been shown that the dynamo in a steady flow and in a time periodic flow are very similar. Further, the results of Ref. 3 do not show any appreciable difference in the dynamo for the ABC flow between the cases when stagnation points exist and when they do not. While we're on the subject of stagnation points, I've heard it asked whether the dynamo is "due to" the stagnation points or "due to" the chaos. The answer is, I believe, contained in my characterization of the fast dynamo in the previous paragraph: the vector stretching is due to the (unstable) stagnation points. The global, and hence fast nature of the dynamo is due to the chaos, which is itself intimately tied up (literally) with the stagnation points. The same relations hold when the chaos (and the dynamo) is associated with unstable periodic orbits.

Specifically, consider the flux through a macroscopic area S which intersects the unstable manifold of a stagnation point $\boldsymbol{p}$. It can be shown that this flux is not dominated by the contribution near the unstable manifold of $\boldsymbol{p}$, whether $\boldsymbol{p}$ is a type A point [eigenvalues of $\partial v_i/\partial x_j$ (-,-,+) and hence with a one-dimensional unstable manifold] or a type B point [eigenvalues (-,+,+) and hence with a two-dimensional unstable manifold.] The relevance of the flux through the macroscopic area S lies in the conjecture of Refs. 4,5, and is briefly discussed three paragraphs down.

If the variational equation (4) is integrated over time intervals $0 < t < T$, $T < t < 2T, \ldots$ with T greater than a correlation time for the chaotic orbit, the variational vectors can be thought of as being related by random matrices

$$\delta \boldsymbol{x}(T) = \boldsymbol{M}_1 \delta \boldsymbol{x}(0), \tag{7}$$

$$\delta \boldsymbol{x}(nT) = \boldsymbol{M}_n \boldsymbol{M}_{n-1} \cdots \boldsymbol{M}_1 \delta \boldsymbol{x}(0). \tag{8}$$

If we put aside the vector nature of the problem for a moment, we can say that $\delta \boldsymbol{x}$ (or the magnetic field if η is negligible) behaves as a product of random numbers, $\delta x_n \sim M_n M_{n-1} \cdots M_1 \delta x_0$ and thus has a lognormal distribution, by the central limit theorem. (That is, $L = \ln|\delta x_n| \sim \sum_i \ln|M_i|$ has a normal distribution.) The mean of L, $<L> = n\lambda_L$ [cf. Eq. (6) and note: $T=1$ for convenience] and therefore the Lyapunov exponent λ_L, can be computed accurately using this relationship. However, if we attempt to compute the mean of $|\delta x(t)|^p = e^{pL}$, the central limit theorem is not accurate enough, because we need to know the distribution far out in the tail. For example, taking the distribution of L to be normal,

$$f(L) = e^{-(L-L_0)^2/\sigma^2}, \tag{9}$$

we compute

$$<L> = L_0, \tag{10}$$

but

$$<e^{pL}> \propto \int e^{pL} e^{-(L-L_0)^2/2\sigma^2} dL$$

$$= e^{pL_0 + p^2\sigma^2/2} \int e^{-(L-L_0-p\sigma^2)^2/2\sigma^2} dL, \tag{11}$$

showing the dominant contribution to be near $L = L_0 + p\sigma^2$, which can be out of the range $|L - L_0| \lesssim \sigma$ where the central limit theorem is valid. (This suggests an answer $<\exp(pL)> = \exp(pL_0 + p^2\sigma^2/2)$, which is valid only in the limit $p \to 0$.)

A refinement can be made which allows accurate computation of moments $<e^{pL}>$. Because $f(L)$ is a series of repeated convolutions of $f_0(\ell)$, where $\ell = \ln|M|$, in Fourier space we have

$$\hat{f}(k) = \hat{f}_0(k)^n$$
$$= e^{-nH(k)}, \tag{12}$$

where we have written $\hat{f}_0(k) = e^{-H(k)}$. We can evaluate

$$f(L) = \int e^{-nH(k) + ikL} dk \tag{13}$$

by steepest descent to find

$$f(L) = \left(\frac{G''(\lambda_L)}{2\pi n}\right)^{1/2} e^{-nG(L/n)}. \tag{14}$$

This has been derived in more detail in Ref. 8 and is related to the distribution function of Lyapunov exponents[13] $\lambda \equiv L/n$.

The reason for considering the moment $<e^L>$ (i.e., for $p=1$) is that it gives the flux $\int \boldsymbol{B} \cdot \hat{\boldsymbol{n}} dS \propto \int |\delta x| dS$, when the averaging is performed over initial conditions. According to the conjecture of Refs. 4,5, the growth rate of the flux through a macroscopic area gives the growth rate of the fastest growing dynamo mode because the flux diffuses over length scales $|\Delta x| \lesssim \eta^{1/2}$. (This conjecture was verified for a class of model flows in Ref. 7.)

The Lyapunov exponent $\lambda_L = <L>/n$ gives a measure of the growth of δx. However, the quantity $\lambda_T = \ln <e^L>/n$ (i.e., for $p=1$) gives the actual growth of the dynamo mode. (Recall that we are ignoring certain aspects of the vector nature at this point, namely the possibility of cancellation.) The quantity λ_T is called the topological entropy in the nonlinear dynamics context, and is in fact precisely defined as the growth rate of the length of a line segment of finite initial length. The connection is that the line segment can be split into many little pieces, each so short that the vector along its length δx_i satisfies the variational equation (4), at least for a finite interval of time. Then the length is the sum of δx_i, or the moment $<e^L>$.

It is easily seen that $\lambda_T \geq \lambda_L$, and in fact equality only holds for uniform stretching $f_0(l) = \delta(l-l_0)$. (In fact $\lambda_T > \lambda_L$ is responsible for the fractal nature of $\boldsymbol{B}(\boldsymbol{x})$.[4,5]) Further, have $\lambda(p) \geq \lambda(q)$ if $p > q$, where $\lambda(p) \equiv \ln <e^{pL}>/n$. Again equality holds only for uniform stretching. Also, $\lim_{p \to 0} \lambda(p)$ equals λ_L. Thus, ignoring the cancellation aspect of the vector nature of the problem (which we will be dealing with very shortly, I promise) we find that the growth rate of the dynamo mode is greater than or equal to the Lyapunov exponent.

The neglected aspect of the vector nature of the problem deals with cancellation. (Finally.) Two nearby points in space $\boldsymbol{x}_1$ and $\boldsymbol{x}_2$, have magnetic fields $\boldsymbol{B} = \delta \boldsymbol{x}$ (i.e., we are still using $\eta = 0$) that were convected and stretched over very different paths before arriving at $\boldsymbol{x}_1$ and $\boldsymbol{x}_2$. It turns out that $\delta \boldsymbol{x}(\boldsymbol{x}_1)$ and $\delta \boldsymbol{x}(\boldsymbol{x}_2)$ must be either parallel or antiparallel if $|\boldsymbol{x}_1 - \boldsymbol{x}_2|$ is small enough. However, the antiparallel possibility allows the flux $\int \boldsymbol{B} \cdot \hat{\boldsymbol{n}} dS$ to have cancellation, so that its growth rate γ satisfies $\gamma \leq \lambda_T$, where λ_T, the topological entropy, is the value obtained above.

These two features of the dynamo, namely the lognormal nature of the field (with a large caveat) and the possibility of partial cancellation which reduces the growth rate, are common to all fast dynamos. In particular, we have recently demonstrated the existence of a fast dynamo in a steady flow[6,8] (i.e., with $\partial \boldsymbol{v}/\partial t = 0$.) There, adjustment of the flow to prevent complete cancellation (i.e., to allow some growth) is crucial. In fact, the reason why a dynamo cannot occur in two dimensional time dependent flows [e.g., $\boldsymbol{v}(x,y,t)$] is that perfect cancellation occurs.

The detailed nature of the dynamo magnetic field (lognormal distribution of $|\boldsymbol{B}|$ — with caveat, as mentioned; cancellation on small length scales) may or may not be directly observable. The kinematic dynamo problem is linear, but the full problem becomes nonlinear when the Lorentz force becomes appreciable. In fact, for $R_m \gg 1$, the field can grow to a very large magnitude $[|\delta \boldsymbol{B}(t)|/|\delta \boldsymbol{B}(t=0)| \sim R_m^{1/2}]$ before the length scales decrease to the point $(|\Delta \boldsymbol{x}| \sim R_m^{-1/2})$ where the linear eigenfunction sets up. It is entirely conceivable that nonlinearity sets in through the Lorentz force long before this stage, and resistivity plays an even more minor role. There is one scenario under which the field structure should be visible. This is when magnetic buoyancy is the dominant nonlinear force, so that as soon as large magnetic fields are generated in the solar convection zone, they are buoyed to the photospheric surface. In this regard, I call your attention to the statistical analysis of sunspot areas of Bogdan et al.[14] These 24,000 sunspot areas, measured over the period 1917–1982, were distributed lognormally over several orders of magnitude. This connection with fast dynamo theory is certainly intriguing and deserves further study.

Summarizing, in the kinematic fast dynamo problem the underlying nonlinear dynamics of the flow [i.e., Eqs. (3), (4)] plays a critical role in the behavior of the dynamo field. In fact, the conjecture of Refs. 4,5 completely reduces the problem to the nonlinear dynamics of the flow (and this conjecture has been verified in Ref. 7). The two important facets of the problem are the approximately lognormal distribution of vector lengths, and the presence of partial cancellation. These features may possibly be reflected in the magnetic fields observed on the sun.

ACKNOWLEDGMENTS

I have benefitted greatly from discussions on these topics with Tom Antonsen, Bruce Bayly, Steve Childress, Jim Hanson, Ittai Kan, and Ed Ott. This work was supported by grant NAGW-1346 of the Solar Physics Branch of NASA, and grant NAG5-1356 of Goddard Space Flight Center.

REFERENCES

1. V.I. Arnold, Ya.B. Zeldovich, A.A. Ruzmaikin, and D.D. Sokolov, Sov. Phys. JETP 81, 2050 (1981).
2. V.I. Arnold and E.I. Korkina, Vestn. Mosk. Univ. Mat. Mekh. 3, 43 (1983 (in Russian)).
3. D. Galloway and U. Frisch, Geophys. Astrophys. Fluid Dyn. 36, 53 (1986).
4. J.M. Finn and E. Ott, Phys. Rev. Lett. 60, 760 (1988).
5. J.M. Finn and E. Ott, Phys. Fluids 31, 2992 (1988).
6. J.M. Finn, J.D. Hanson, I. Kan and E. Ott, Phys. Rev. Lett. 62, 2965 (1989).
7. J.M. Finn and E. Ott, Phys. Fluids B 2, 916 (1990).
8. J.M. Finn, J. Hanson, I. Kan, and E. Ott, Phys. Rev. Lett. 62, 2695, (1989); Phys. Fluids B, 3, 1250 (1991).
9. B.J. Bayly, Phys. Rev. Lett. 57, 2800 (1986).
10. B.J. Bayly and S. Childress, Phys. Rev. Lett. 59, 1573,(1987); Geophys. and Astrophys. Fluid Dyn., 44, 211 (1988).
11. S. Childress, Phys. Earth Planet. Int. 20, 172 (1979).
12. F. Perkins and E. Zweibel, Phys. Fluids 30, 1079 (1987).
13. P. Grassberger, R. Badii, and A. Politi, J. Stat. Phys. 51, 135, (1988); H. Hata, T. Horita, H. Mori, T. Morita, and K. Tomita, Prog. Theor. Phys., 80, 809 (1988); E. Ott, C. Grebogi, and J. Yorke, Phys. Lett. A, 135, 343 (1989); E. Ott and T. Antonsen, Phys. Rev. A, 39, 3660 (1989).
14. T. Bogdan, P. Gilman, I. Lerche, and R. Howard, Ap.J. , (Apr. 1, 1988).

UNDERSTANDING THE SOURCE OF THE SOLAR ACTIVITY CYCLE: RESULTS AND PROSPECTS FROM HELIOSEISMOLOGY

Philip R. Goode[†]
Institute of Astronomy, University of Cambridge, Cambridge, England

ABSTRACT

Helioseismic studies have revealed that the only sharp change in the Sun's internal rotation occurs near the interface between the convective zone and the radiative interior. This region is generally regarded as the source of the solar activity cycle. Other helioseismic clues to the properties of the interface concern the magnetic field and the temporal stability of rotation there.

INTRODUCTION

Understanding the Sun's internal rotation and magnetism is critical in our efforts to determine the nature of the engine driving the spectacular phenomena we observe on the Sun's surface and in its atmosphere. The OSL satellite will greatly increase our understanding of these phenomena and further motivate our efforts to understand the solar interior. What we know about the Sun's interior we have learned from helioseismology.

The confirmed solar oscillations have been observed in either intensity fluctuations in white light or Doppler shifts in a particular line. The observations have been of the resolved disk leading to solar five-minute acoustic mode data which can best teach us about the region near the base of the convection zone. See Harvey[1] for a review of the observational approaches and data. The most solid helioseismic inferences are the location of the base of the convection zone, Christensen-Dalsgaard et al.[2], and the internal rotation rate between about $0.5R$ and $0.9R$, Dziembowski, et al.[3]. Roughly stated, surface-like differential rotation persists through the convection zone with an abrupt transition to solid body-like rotation beneath. Knowing the precise abruptness of this transition and the origin of it are central in our efforts to understand activity. After all, it is generally regarded that the source of solar activity lies in this region. If the confluence of angular and radial gradients were the criterion for locating the dynamo, the helioseismological results point to the base of the convection zone. Magnetic buoyancy considerations led dynamo theorists to this location as well, Speigel and Weiss[4]. There is also the seismic hint of a large magnetic field at the base of the convection zone, Dziembowski and Goode[5,6].

Sizeable and systematic frequency changes in the seismic data have been observed through the solar cycle and are associated with the magnetic fields in the active regions, Libbrecht and Woodard[7] and Woodard, et al.[8]. There is also weaker evidence that the Sun's internal rotation changes in the radiative interior over the cycle, Goode and Dziembowski[9].

Our purpose in this review is to discuss the confirmed seismic results about

[†] Permanent address: Department of Physics, New Jersey Institute of Technology, Newark, NJ 07102, USA

the Sun's internal rotation and magnetism, and speculate about the seismic knowledge we can expect to gain in the next few years. Further, we will explore possible ramifications of such knowledge.

THE OSCILLATIONS

An individual acoustic mode, $\boldsymbol{\xi}$, is defined by

$$\boldsymbol{\xi}_{n,l,m}(r,\theta,\phi,t) = r\left[y_{n,l}(r), z_{n,l}(r)\frac{\partial}{\partial\theta}, z_{n,l}(r)\frac{1}{\theta}\frac{\partial}{\partial\phi}\right]Y_l^m(\theta,\phi)\exp(i\omega t) \quad (1)$$

where $y_{n,l}$ and $z_{n,l}(r)$ represent the radial and horizontal displacement of the fluid and ω is the frequency of the oscillation. The n, l and m labels are the radial order, angular degree and azimuthal order of the oscillation, respectively. The solar oscillations appear in (nl)-multiplets. In the absence of rotation and magnetic fields, the multiplets would be $(2l+1)$-fold degenerate in m. In a slowly rotating star like the Sun, the rotation would lift the degeneracy causing a nearly uniformly spaced fine structure in each multiplet. A sizeable magnetic field inside the Sun would alter such a pattern in the fine structure. In our helioseismic inferences, we rely on fine structure data from many multiplets having information from differential samplings of the Sun's interior.

In general, stellar oscillations are non-radial and non-adiabatic in nature. We calculate the properties of these oscillations by perturbing and solving the equations of continuity, hydrostatic equilibrium, conservation of energy and a subsidiary condition relating the radiative flux and the mean radiative intensity, see Unno, et al.[10] for a review. The well-confirmed observed solar oscillations have a period of about five minutes and range in degree between $l \sim 10$ and $l \sim 100$. For these oscillations we can usefully study their properties by studying the single asymptotic equation,

$$f'' + f\left[\frac{\omega^2 r^2}{c^2} - l(l+1) + O(|\frac{d\ln\rho}{d\ln r}|^2, 1)\right] = 0 \quad (2)$$

for the oscillations in their high-frequency limit, where ρ is the density and c is the speed of sound. The inner turning point for these sound waves occurs when the first two terms in brackets are comparable. Thus, the lower the degree of the oscillation the more deeply it samples. The seismology proceeds on the differential sampling of the interior that this implies. The lowest degree oscillations sample all the way to the Sun's center. Near the surface the reflection point for these global modes is reached when the square of the derivative of the density becomes comparable to the first term in brackets. Since the well-confirmed solar oscillations have a period near five minutes, they all tend to see the region near the surface in roughly the same l-independent way. Thus we have the result that at this point in the history of helioseismology, we have learned the most about the region near the base of the convection zone.

The solar oscillation data are usually described by

$$\nu_{n,l,m} - \nu_{n,l,0} = L\sum_{i=1}^{N} a_{i,n,l}P_i(\frac{m}{L}) = \sum_{i=1}^{N} \alpha_{i,n,l}P_i(\frac{m}{L}) \quad (3)$$

where $\nu_{n,l,m}$ is the frequency a particular oscillation and P_i is a Legendre polynomial and $L = l$ or $\sqrt{l(l+1)}$ depending on the choice of the observer and $N = 5$ or 6 for now. The fact that this simple expansion fits the data so well implies that the Sun rotates on a single axis and that perturbations manifest in the splittings have the rotation axis as their axis of symmetry. For instance, an intense inclined magnetic field would induce an unsteady perturbation which would result in observers in the Earth's frame reporting at least $(2l+1)^2$-peaks in each affected (nl)-multiplet, Dicke[11]. The a/s in equation (3) are the splitting coefficients used in the seismic inferences. The a/s are sometimes used in conjunction with the symmetric part(with respect to m) of the splittings. The odd a/s – a_1, a_3 and a_5 – are uniquely associated with the linear effect of rotation and correspond to a rotation law of the form

$$\Omega(r,\theta) = \Omega_0(r) + \Omega_1(r)\cos^2\theta + \Omega_2(r)\cos^4\theta. \tag{4}$$

This rotation law has the same structural form as that observed for the solar surface. The even-a/s or α/s are not straightforwardly associated with any perturbation beyond the second order effects of rotation-like distortion. However, it is generally regarded that a large part of the symmetric signal arises from the effect of magnetic fields in the active latitudes where the perturbation would be fairly l-independent as discussed in conjunction with equation (2). Thus, the α/s are introduced because near-surface perturbations, like centrifugal distortion or magnetic fields in the active latitudes, induce splittings which tend to be independent of l.

THE INTERNAL ROTATION

The first helioseismic inference was the rotation rate in the Sun's equatorial plane, Duvall, et al.[12]. They determined that the rate was quite close to the observed surface equatorial rate with some tendancy to decline going inwards. Subsequently, Duvall, Harvey and Pomerantz[13] observed that the convection zone mimics the surface differential rotation of the Sun. Then, Brown, et al.[14] and Dziembowski, et al.[3] reported that near the base of the convection zone there is an abrupt transition to solid-body-like rotation. These conclusions follow from solving the inverse problem for rotation. To see the origin of these results for rotation, we pose a simplified, but reasonably accurate inverse problem which, with the aid of a mean rotation law, enables us also to roughly perform the inversions by eye.

The inverse problem for the determination of $\Omega_0(r)$, $\Omega_1(r)$ and $\Omega_2(r)$ involves three coupled equations. At the current level in the accuracy of the five-minute solar oscillation data, we may safely ignore the coupling and for our purposes pretend that $l \gg 1$. To a useful approximation, the inverse problem then becomes

$$\frac{1}{2\pi} \int K_{n,l}(r)\Omega_0(r) \approx a_{1,n,l} + a_{3,n,l} + a_{5,n,l} \tag{5}$$

and

$$\frac{1}{2\pi} \int K_{n,l}(r)\Omega_1(r) \approx 5a_{3,n,l} - 14a_{5,n,l} \tag{6}$$

and
$$\frac{1}{2\pi}\int K_{n,l}(r)\Omega_2(r) \approx 21 a_{5,n,l}. \tag{7}$$

The splitting kernel or sampling function $K_{n,l}$ depends on the eigenfunction of equation (1) which is determined using a solar model in the solution of the dynamical equations following from perturbing the equilibrium state of the model Sun. The kernel is given by

$$K_{n,l}(r) = \left[y_{n,l}^2(r) + l^2 z_{n,l}^2(r) - 2y_{n,l}(r)z_{n,l}(r)\right]\rho r^4. \tag{8}$$

As mentioned, the well-confirmed data ranges in l from about 10 to 100. Of course the basic observational approach is to gather as long a datastring as possible with the maximum possible coverage each day. After all, temporal sidebands are a major obstacle in the data reduction and mode identification. The reduction process is somewhat eased if one averages the data over n; however, the price is reduced accuracy in the inversions. The best data to date are from Big Bear Observatory where the sets are roughly 100 days long and n is not averaged over, Libbrecht and Woodard(ref. 7 and private communication). The Big Bear datasets confirm earlier sets, averaged over n, from Duvall, Harvey and Pomerantz[13], Brown and Morrow[15] and Rhodes, et al.[16]. The summer of 1986 Big Bear data for a_1, a_3 and a_5 are shown in figure 1. To make these data easier to look at in our eyeball inversion, we have averaged them over n and further averaged them into bins 5 l-values wide. Using equations (5)–(7), we have determined a mean rotation law from the data in figure 1. It is

$$\frac{1}{2\pi}\bar{\Omega}(\theta) = 460.2 \pm 0.2 - (58.3 \pm 1.3)\cos^2\theta - (73.1 \pm 2.6)\cos^4\theta \quad \text{(nHz)}. \tag{9}$$

Recalling the observations of Duvall, Harvey and Pomerantz[13], it is not surprising that this mean rotation law is quite close to the observed differential rotation for the solar surface.

From figure 1, it is clear that we know some of binned mean a_1's to better than 1% and that the mean rotation describes this part of the data. If one employed the prediction of standard dynamo theories that the rotation through the convection zone is constant on cylinders, then the calculated a_1's would be about 420nHz for l-values near 40 – a $\sim 10\sigma$ discrepancy. We do not know the a_3's as well as the a_1's. The observed and calculated a_3's diverge for low l-values – this is the source of the calculated transition away from surface-like differential rotation toward solid-body-like rotation deep in the convection zone. Our knowledge of the a_5's is even poorer. Thus, the higher the latitude in the interior, the poorer our knowledge is of the internal rotation.

A question of paramount importance is how abrupt is the change near the base of the convection zone. We can determine the granularity of the information we can extract from the oscillation data about the rotation. The resolution is shown in figure 2 for the region at the base of the convection zone for the aforementioned Big Bear data. Dziembowski, et al.[3] have shown that it is about the same width over which the change from differential rotation to solid body rotation occurs. Thus the true transition could be abrupt. With this in mind,

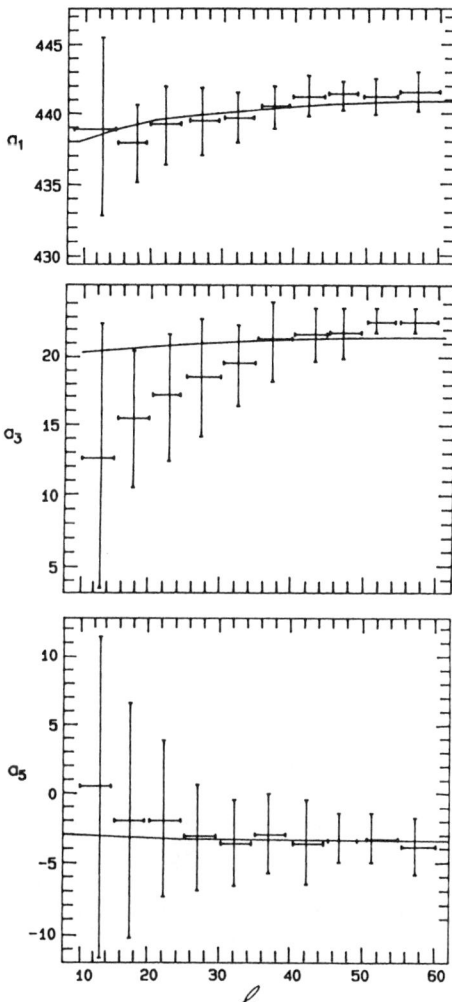

Fig. 1. Odd a-splitting coefficients(nHz) vs. l from the data of Libbrecht and Woodard[7]. The solid line represents the determination of the coefficients from the mean rotation law of equation (9).

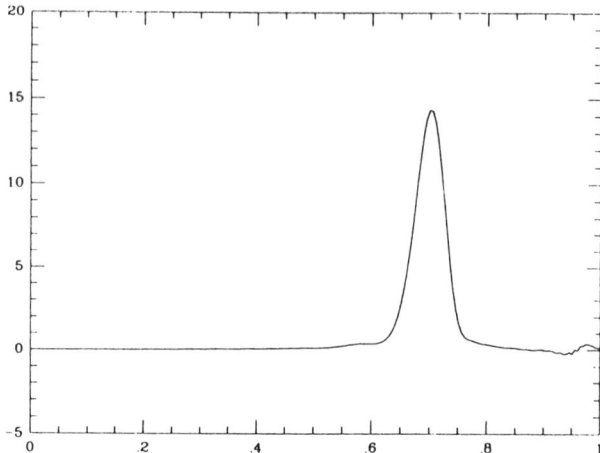

Fig. 2. Resolution kernel for $0.7R$ vs. fractional radius using data of Libbrecht and Woodard[7].

we solve the fully-coupled inverse problem using the Big Bear data but allowing a single discontinuity of arbitrary size to occur at various radii. It turns out that χ^2 per degree of freedom attains its minimal value when the discontinuity is at the base of the convection zone. The result is shown in figure 3, where one sees surface-like differential rotation through the convection zone and solid body rotation beneath. The solid body rate corresponds to that at an intermediate latitude on the surface. The implied net angular momentum flow would still be toward the surface.

In the future, the Global Oscillations Network Group(GONG) will provide long datastrings without the usual day-night gaps from a series of continuously operating telescopes placed at various longitudes around the world. With such data, we will be able to ascertain the true nature of the interface region which is so critical in our efforts to understand the underlying source of the activity cycle.

CYCLE DEPENDENCE OF THE DATA AND INTERNAL MAGNETISM

Kuhn[17] has shown that the symmetric part of the splitting data, the even a/s or α/s, vary systematically through the activity cycle. Further, he showed that there is a corresponding variation of the cool latitudinal temperature bands observed by Kuhn, Libbrecht and Dicke[18]. Libbrecht and Woodard[7] have shown that these bands may be regarded as a proxy for surface activity which also causes the time dependence of the splittings.

To understand the nature of such a near-surface perturbation, we define a general second order change in the frequency, $\Delta\omega$ due to some operator $\Delta \boldsymbol{L}$,

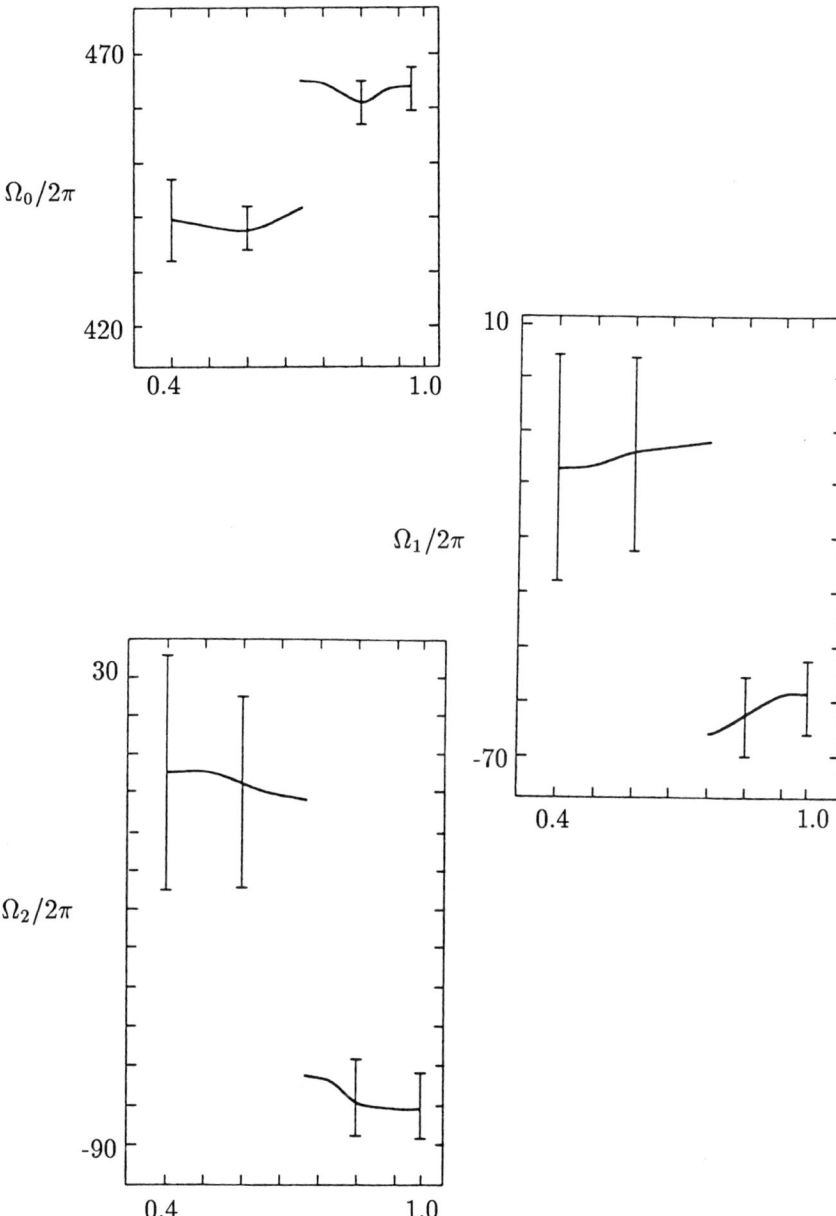

Fig. 3. $\Omega_0/2\pi$, $\Omega_1/2\pi$ and $\Omega_2/2\pi$ in mHz vs. fractional radius from the data of Libbrecht and Woodard[7] allowing a discontinuity at the base of the convection zone

where
$$\Delta\omega = \frac{\int \boldsymbol{\xi}^* \cdot \Delta \boldsymbol{L}(\boldsymbol{\xi})\rho d^3 r}{2\omega \int \boldsymbol{\xi}^* \cdot \boldsymbol{\xi}\rho d^3 r}. \tag{10}$$

We presume that the angular part of $\Delta \boldsymbol{L}$ is given by an expansion in $P_{2s}(\cos\theta)$ and that $l \gg s$. This is a good approximation for us where we have $l > 10$ and $s \leq 3$. Goode and Kuhn[19] have shown that in this limit, the angular integral in the numerator of equation (10), $\tilde{Q}$, becomes

$$\tilde{Q}_{s,m,l} = (-)^s \frac{(2s-1)!!}{(2s)!!} P_{2s}(\frac{m}{L}). \tag{11}$$

Assuming $\Delta \boldsymbol{L}$ is a near-surface perturbation, then the part of $\boldsymbol{\xi}(r)$, where $\Delta \boldsymbol{L}$ is large, is very nearly radial and the Y_l^m's carry all the l-dependence of the eigenfunction(this follows from the nature of the outer turning point as discussed for equation (2)). Using this property of $\boldsymbol{\xi}(r)$ and comparing equation (3) with equations (10) and (11), equation (10) may be re-phrased as

$$\alpha_{2s} \propto \frac{\gamma_s(\nu)}{I_{n,l}}, \tag{12}$$

where $I_{n,l}$ is called the mode inertia or mode mass

$$I_{n,l} = \int \left[y_{n,l}^2 + l(l+1)z_{n,l}^2 \right] \rho r^4 dr. \tag{13}$$

The mode mass is a strong function of frequency varying by about two orders of magnitude between 2mHz and 4mHz which is roughly the central range of the five-minute oscillation band. The frequency dependence of $\gamma_s(\nu)$ is sensitive to how the eigenfunctions are normalized. If we normalize them at the photosphere, the $\gamma_s(\nu)$'s are weak functions of frequency and most of the frequency dependence in equation (12) comes from the denominator. This behavior in the even-a's has been reported by Libbrecht and Woodard[7] when they consider the differences between their 1986 and 1988 data. Thus, the general considerations in our evaluation of equation (10) re-enforce the conclusion of Libbrecht and Woodard[7] that the time-dependence in their data is due to near-surface perturbation. The cool latitudinal temperature bands of Kuhn, Libbrecht and Dicke[18] can be associated with the γ_s's, and therefore be translated into a_{2s}'s, equation (12). The resulting a_2's and a_4's have been calculated by Goode and Kuhn[19] both from the temperature or activity data(called A in figure 4) and from the oscillation data. For the oscillation data, they removed the effect of centrifugal distortion from the even-a's. The two kinds of a's are compared as a function of time in figure 4. The a_2's reached a minimum near the activity minimum in 1986 while the a_4's have continued to decline. The a's at the various timepoints are taken from the observational data of Duvall, Harvey and Pomerantz[13], Brown and Morrow[15], Rhodes, et al.[16], Jefferies, et al.[20] and Libbrecht and Woodard[7]. The consistency in the systematic variation of the data, using data from various observers, represents an impressive endorsement of the reliability of oscillation

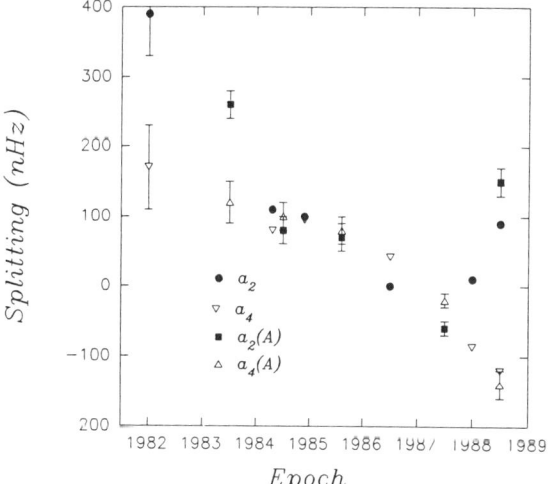

Fig. 4. The a/s from oscillation data and activity vs. time.

data. Still, it would be somewhat depressing if we could only learn about the near-surface region from the symmetric part of the oscillation data. This raises the question of a second signal in this part of the data; a second signal which tells something about the deeper interior.

If there were a second signal in the data, what would we expect it's source to be? The available data contains its most accurate information about the region near the base of the convection zone. In that region, one would expect that the shearing action of differential rotation could stretch even a minuscule poloidal magnetic field into a sizeable quadrupole toroidal magnetic field. The best data in which to look for such a signal would be the Big Bear data of 1986 and 1988 because they are not averaged over n. This averaging in the other sets probably precludes seeing any secondary signal in them. Dziembowski and Goode[5,6] have formulated the inverse problem for a toroidal field. This formulation is built on generalizing equation (12), where

$$a_{2j,d} = \sum_{k \geq j} \int A_{k,j,d} \beta_k(r) dr + (a_{2j,d})_{\rm rot} + \frac{I_* \gamma_j(\nu_d)}{I_d L_d}. \tag{14}$$

$A_{k,j,d}$ is the magnetic field kernel given in equations (26) and (45) of Dziembowski and Goode[5], d is the mode counter and $(a_{2j,d})_{\rm rot}$ is the second order effect of rotation calculated using the rotation law following from the anti-symmetric part of the data, Dziembowski and Goode[21]. The I_* is a normalization factor introduced to make the γ/s comparable in magnitude to the a/s in the data. The β/s are the measure of the toroidal field and are defined by,

$$B_\phi^2 = 4\pi p \sin^2 \theta \sum_{k=1} \beta_k(r) \cos^{2k-2} \theta, \tag{15}$$

where p is the local gas pressure. Simultaneous inversions for β_3 and γ_3 and then for β_2 and γ_2 have been performed by Dziembowski and Goode[6]. Their procedure was to use the a_6 from a dataset to solve for β_3 and γ_3 and then use that result and the a_4 data to solve for β_2, the quadrupole toroidal field, and γ_2. The results are shown in figures 5 and 6 respectively.

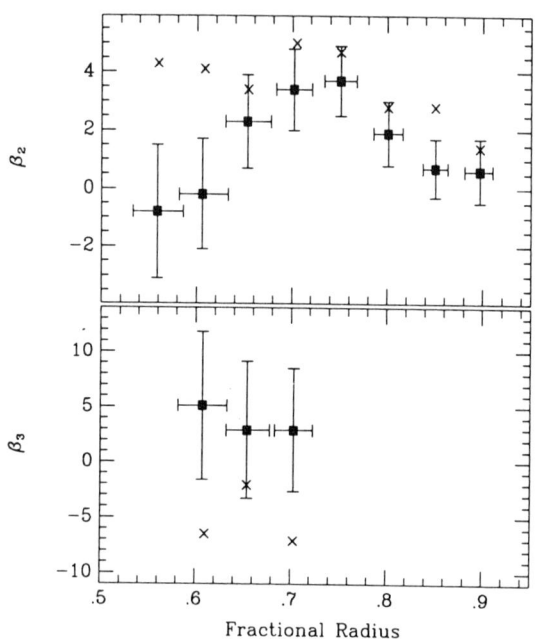

Fig. 5. The toroidal magnetic field measures β_2 and β_3. The boxes(crosses) represent the results following from the 1986(1988) data. The vertical error bars show the rms errors. The horizontal ones represent the resolution. The corresponding error bars for the 1988 data are similar.

The toroidal field sampling functions, $A_{k,j,d}$, are stronger functions of n and l than are their counterparts for rotation. Thus, the granularity in our determination of the field at the base of the convection zone is no worse than that for rotation as shown in figure 3.

From both the 1986 and 1988 datasets of Libbrecht and Woodard[7], β_3 values could only be determined in the region near the base of the convection zone and the results are consistent with zero. We see no secondary signal in the a_6 data. The only statistically significant results for β_2 occur near the base of the convection zone where both datasets yield something like a steady 3-4σ effect. This corresponds to a steady megagauss quadrupole toroidal field. Such a field is considerably larger than one would expect from standard dynamo theories. The magnetic pressure caused by this field would be about three orders of magnitude weaker than the local gas pressure. Roughly speaking the field would seem to

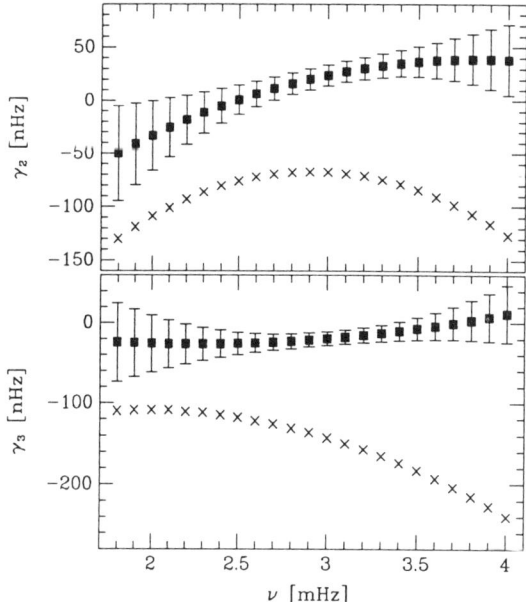

Fig. 6. The $\gamma(\nu)$-functions describing the near-surface effect in symmetric splittings. Boxes (crosses) represent the γ's from the 1986 (1988) data. The least-square error bars are about the same for the corresponding points from each dataset.

be spread over about 0.1R, and therefore contains energy comparable to the luminosity of the Sun through an entire activity cycle. We remark that if such a field were to observably change between 1986 and 1988, to keep the luminosity constant, rotation would have to have changed much more than seismic data imply. We conclude that there is a steady secondary signal in the symmetric part of the seismic data which can be extracted from the strongly varying symmetric data. The variation in the data is clear in figure 6. In 1986, at solar minimum, the γ's were consistent with zero. In 1988, the values were significantly different from zero. As shown by Libbrecht and Woodard[7] and Woodard, et al.[8], this difference is due to the near-surface perturbation caused by the magnetic fields in the active regions in 1988. With the GONG data, we will be able to determine, once and for all, whether or not there is a sizeable magnetic field near the base of the convection zone.

The reported time dependence in the solar oscillation data arises in the symmetric part of that data. Of course this makes one question whether there is a second, smaller time-dependence in the antisymmetric part of the data. This is really asking the question – does the Sun's internal rotation depend on time?

SOLAR CYCLE DEPENDENCE OF THE INTERNAL ROTATION?

To look for a time-dependence in the seismically determined internal rotation, it is wisest to examine that rate in the Sun's equatorial plane where we best know it. We first compare the 1986 and 1988 Big Bear datasets because they were gathered and reduced in the same way. In figure 7, we plot the difference between $a_1 + a_3 + a_5$ between 1988 and 1986 where the data are binned like those in figure 1. From equation (5), one would conclude that this difference is a measure of the change in rotation in the equatorial plane. The difference is consistent with zero except for low l-values. This behavior would seem to imply that the only time-dependence occurs in the deep interior. On the other hand, one might try to imagine that this difference is due to a near-surface perturbation instead. However, it is difficult to conceive a way in which sunspots could preferentially alter low-l splittings in the antisymmetric part of the data.

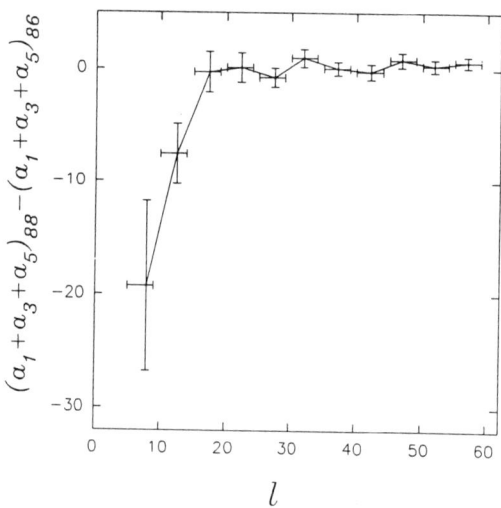

Fig. 7. The difference between the measure of rotation in the equatorial plane for the years 1988 and 1986 vs. l. The measure is defined in equation (5).

The two rotation laws for the equatorial plane were calculated by Goode and Dziembowski[9] and are shown in figure 8. The rotation rate is most steady at the base of the convection zone where we know it best. The most significant change in rotation($\sim 1.5\sigma$) occurs in the deep interior with that rate being larger at solar activity minimum. One could argue that this difference is a reflection of the true observational errors in the data. This may well be true, but taking the result and its errors at face value, our confidence in its veracity would be strengthened if consistent changes occured in other datasets – namely, those used in figure 4. When the rotation is calculated in the same way for all the datasets, Goode and Dziembowski[9] determined that the rate at the base of the convection zone shows essentially no variation with the cycle. Whereas, the rate deeper down, as shown in figure 9, changes in a way which is anti-correlated with activity.

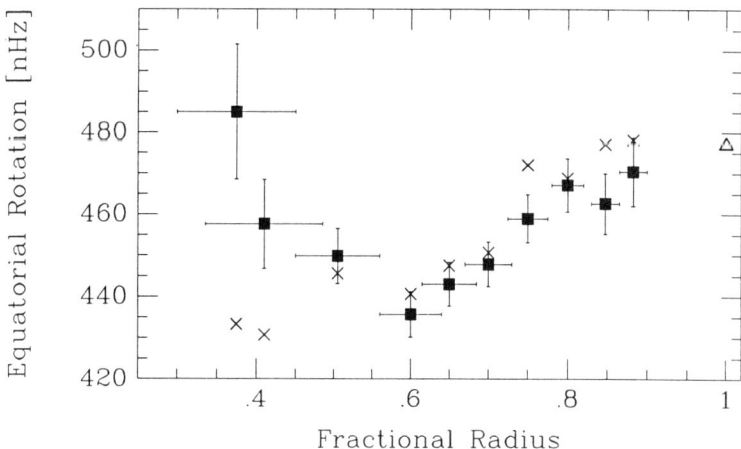

Fig. 8. Two internal rotation laws for the equatorial plane from the 1986 and 1988 data of Libbrecht and Woodard[7] The error bars are as defined in figure 5.

Considering the thermal timescale in the radiative interior and that the luminosity of the Sun is observed to be nearly constant, the only way the rotation in the radiative interior could change on such a short timescale is if there were a nearly adiabatic exchange of energy between rotation and magnetism – a torsional oscillation.

The possibility of a torsional oscillation inside the Sun was first put forward by Walen[22]. What he had in mind was a steady roughly kilogauss poloidal field permeating the inside of the Sun and together with differential rotation it forces an exchange of energy between an induced toroidal magnetic and a time-dependent part of the internal rotation. The period of this torsional oscillation would depend primarily on the amplitude of the poloidal field. A kilogauss field is necessary to make the 22-year activity cycle period. In those days the Sun was believed to have a convective core, and Walen argued that the torsional oscillation was convectively driven from the core. This view of activity has many problems, not the least of which would be how a steady kilogauss poloidal field is maintained in the convection zone. This picture was supplanted in the 1960's by distributed dynamo models. Problems associated with magnetic buoyancy led to the idea that the seat of the dynamo was, instead, near the base of the convection zone, Spiegel and Weiss[4]. At that time, it probably became worthwhile to consider a torsional oscillation which is confined to the radiative interior as an alternative or supplement to a deeply seated dynamo. Now, the helioseismic data further motivates us to consider a torsional oscillation in the radiative interior.

The equations describing the purely horizontal torsional oscillation were

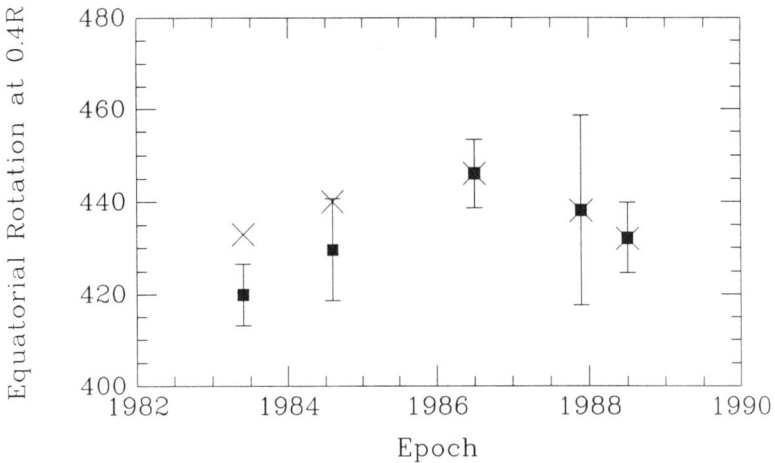

Fig. 9. The rotation rate in the equatorial plane at 0.4R. The X's are the inverse of the mean sunspot number calibrated to the rates from the 1986 and 1988 data of Libbrecht and Woodard[7].

given by Cowling[23]. The equation of motion is

$$\rho r^2 \sin^2\theta \frac{\partial^2 \Omega}{\partial t^2} = \frac{1}{4\pi}(\boldsymbol{B}_p \cdot \boldsymbol{\nabla})\left[r^2 \sin^2\theta(\boldsymbol{B}_p \cdot \boldsymbol{\nabla})\Omega\right]. \tag{16}$$

The magnetic induction equation is

$$\frac{\partial B_\phi}{\partial t} = r\sin\theta(\boldsymbol{B}_p \cdot \boldsymbol{\nabla})\Omega. \tag{17}$$

The helioseismic data imply that the outer part of the radiative interior rotates like a solid body with, perhaps, a small time dependent part deeper down. This suggests that the kilogauss poloidal field exists in the outer part of the radiative interior where it stiffens the matter so that there is no cycle dependence in the rotation there. Presumably, deeper down this field is weaker and that is where the torsional oscillation occurs. Of course such an oscillation would generate meridianal circulations and induced poloidal fields. However, the adiabatic nature of a 22-year oscillation in the radiative interior implies that the motion is very nearly horizontal. The toroidal field generated by this oscillation has the requisite ~10KG size.

Of course, this helioseismically motivated suggestion has many unclear points primarily concerning how the torsional oscillation would be driven and the resulting fields would escape the interior to become the spectacular phenomena observed on the solar surface. Forthcoming oscillation data will tell us whether or not there is a torsional oscillation in the radiative interior.

ACKNOWLEDGEMENTS

P.R.G. is partially supported by AFOSR-89-0048.

REFERENCES

1. J.W. Harvey, Seismology of the Sun and Sun-like Stars (ESA, 1989), p. 1.
2. J. Christensen-Dalsgaard, T.L. Duvall, Jr., D.O. Gough, J.W. Harvey and E.J. Rhodes, Jr., Nature 315, 378 (1985).
3. W.A. Dziembowski, P.R. Goode, and K.G. Libbrecht, Ap.J.Lett 337, L53 (1989).
4. E.A. Spiegel and N.O. Weiss, Nature 287, 616 (1980).
5. W.A. Dziembowski and P.R. Goode, Ap.J. 347, 540 (1989).
6. W.A. Dziembowski and P.R. Goode, Ap.J. , in press (1991).
7. K.G. Libbrecht and M.F. Woodard, Nature 345, 779 (1990).
8. P.R. Goode and W.A. Dziembowski, Nature 349, 223 (1991).
9. M.F. Woodard, J.R. Kuhn, N. Murray and K.G. Libbrecht, Ap.J.Lett. 373, L81 (1991).
10. W. Unno, Y. Osaki, H. Ando, H. Saio and H. Sibahashi, Nonradial Oscillations of Stars (University of Tokyo Press, 1989).
11. R.H. Dicke, Nature 300, 693 (1982).
12. T.L. Duvall, W.A. Dziembowski, P.R. Goode, D.O. Gough, J.W. Harvey and J.L. Leibacher, Nature 310, 22 (1984).
13. T.L. Duvall, Jr., J.W. Harvey and M.A. Pomerantz, Nature 321, 500 (1986).
14. T.M. Brown, J. Christensen-Dalsgaard, W.A. Dziembowski, P.R. Goode, and C.A. Morrow, Ap.J 347, 540 (1989).
15. T.M. Brown and C.A. Morrow, Ap.J.Lett. 314, L21 (1987).
16. E.J. Rhodes, Jr., A. Cacciani, S. Korzennik, S. Tomczyk, R.K. Ulrich and M.F. Woodard, Ap.J. 356, 310 (1990).
17. J.R. Kuhn, Ap.J.Lett. 331, L131 (1988).
18. J.R. Kuhn, K.G. Libbrecht and R.H. Dicke, Science 242, 908 (1988).
19. P.R. Goode and J.R. Kuhn, Ap.J. 356, 310 (1990).
20. S.M. Jefferies, M.A. Pomerantz, T.L. Duvall, Jr., J.W. Harvey and D.B. Jaksha, Seismology of the Sun and Sun-like Stars (ESA, 1988), p. 279.
21. W.A. Dziembowski and P.R. Goode, Ap.J. , submitted (1991).
22. C. Walen, Ark.Mat.astr.Fys. A33, 1 (1946).
23. T.G. Cowling, Magnetohydrodynamics (Interscience, 1957).

ALTERNATIVE CORONAL HEATING MECHANISMS

R.N. Sudan
Lab. of Plasma Studies, Cornell University, Ithaca, NY 14853

D.W. Longcope
Lab. of Plasma Studies, Cornell University, Ithaca, NY 14853

ABSTRACT

A unified treatment of the dynamics of the random twisting and untwisting of a solar magnetic loop by photospheric motion is presented. For motions fast compared to Alfvén transit time, the shear Alfvén waves damp rapidly on the stochastic field lines of the loop. For slow motions, the loop passes through a sequence of quasi-static equilibria but the response is also punctuated by impulsive events identified as nanoflares. The statistics of these events indicate a Poisson process; the frequency of these events scales as ΔE_M^{-1} where ΔE_M is the energy in each event. Their contribution to heating is estimated.

INTRODUCTION

The general nature of the physical mechanisms responsible for solar coronal heating are widely accepted but controversy settles on the specific details. The source of the mechanical energy required for heating the corona clearly originates in photospheric motions and perhaps, even deeper. The magnetic field plays a crucial role in the transport of this energy to the corona. For photospheric motions, whose time scale is much less than the transit time $\tau = L/v_A$ of the shear Alfvén waves (L is loop length, v_A is the Alfvén velocity), the excitation energy propagates as a shear Alfvén wave. It is well known that the compressional and slow Alfvén waves are unable to propagate beyond the chromosphere. For low frequency photospheric motions $\omega\tau \ll 1$, Alfvén waves will not be excited but the magnetic flux tubes of the loop will be slowly twisted and the time evolution is quasi-static in nature. As a result of this twisting, currents flow parallel and antiparallel to the magnetic field but with very small net current and energy is stored in the magnetic loop. The central issues are (i) the rate of dissipation of the energy in Alfvén wave motion and (ii) the rate of release of the stored magnetic energy in the loop.

Eventually the dissipation of wave energy or stored magnetic energy is governed by the coefficients of dissipation viz., the resistivity and viscosity. Because the Lundquist number $R_M = 4\pi\, a\, v_A/\eta \gtrsim 10^{10}$ is very large for relevant transverse scale size a of a typical loop, the time scale for such dissipation is orders of magnitude larger than needed to relate to the observed power requirements of ~ 1 watt/cm^2 [1]. In order to reconcile the physical mechanism of dissipation with observations, it has been argued that the relevant scale size a has to be drastically reduced, so that the effective $R_{M*} \sim 1$. Furthermore, this reduction must be self-consistently demonstrated on theoretical grounds.

ALFVÉN WAVE HYPOTHESIS

In a uniform magnetic field the damping length of a shear Alfvén wave $l_d \sim R_M \lambda_\|$ where $\lambda_\|$ is its wavelength in the direction of propagation. This is orders of magnitude too large. In an inhomogeneous magnetic field it has been demonstrated by many authors[2] that $l_d \sim R_M^{1/3} \lambda_\|$. The damping length is reduced because inhomogeneity of the magnetic field causes spatial variations in the Alfvén velocity leading to wavefront filamentation. This filamentation reduces the transverse scale size of the wavefront originally around $\sim a$, to $a_*/a \sim R_M^{-2/3}$. However, Hollweg[3] has maintained that $l_d/\lambda_\| < 50$ if Alfvén wave dissipation is indeed a factor in coronal heating. Since $R_M \sim 10^{10}$ obviously Hollweg's criterion is not satisfied. Before rejecting the Alfvén wave hypothesis it is well to understand the fine structure of the magnetic field of a coronal loop. The smoothness of the magnetic field lines is determined by the quantity

$$\alpha = \frac{1}{z} \sum_j \ln \frac{|\delta \boldsymbol{x}_\perp^j(z)|^2}{|\delta \boldsymbol{x}_\perp^j(0)|^2} \quad , \tag{1}$$

where $\delta \boldsymbol{x}_\perp$ is the distance between two neighbouring field lines. The summation is over a large bundle of field lines contained, at $z = 0$, within an element of area $d^2 S$. If α asymptotes to zero as $z \to \infty$ then the field lines stay close to each other and the magnetic field is laminar. However, if α is a positive constant (Liapounov exponent) as $z \to \infty$, neighbouring field lines locally separate at an exponential rate from each other. Such a magnetic field is stochastic. The wavefronts of an Alfvén wave propagating along a stochastic field filaments at a ferocious rate because each element of the phase front $d^2 S$ is stretched and distorted at an exponential rate determined by α but conserves the area $d^2 S$. The transverse fine structure generated in the wave front causes it to damp rapidly with a damping length[4] given by $l_d \sim \lambda_\| \ln R_M$. In a finite length coronal loop the maximum value of z is L and the criterion for stochasticity, strictly speaking, is not satisfied. If $l_d < L$ the loop length, then it does not really matter that z is limited to L in the expression for α. Detailed calculations[5,6] show that for reasonable configuration of parallel and antiparallel current structures in a loop, a significant fraction of the loop volume could have stochastic field lines. Thus, dissipation of the energy in Alfvén waves can take place over acceptable distances.

The source of such energy is in the photosphere. Propagation studies[7] show that the efficiency of propagation of Alfvén energy from photospheric level to the corona survives the rapid expansion of the magnetic field in the chromosphere with a transmission efficiency in the range of 10% for the frequency range of interest. However, Parker[8] points out that loops of varying sizes appear equally bright in x-rays. Since, the minimum frequency for Alfvén wave excitation varies inversely as the loop length it follows that the available power source in photospheric motion must be flat with frequency, if the loops are equally bright. Since most of the energy content in photospheric motion is presumably in the larger, slower eddies Parker's objection to Alfvén wave heating is now limited to the energy source rather than to the lack of appropriate dissipative processes[8]. We return to this point in a later section.

QUASI-STATIC EVOLUTION OF CORONAL MAGNETIC FIELDS

If it is true that there is no source of energy available to trigger Alfvén waves at the appropriate frequencies, the only remaining mechanism is the slow twisting of the flux tube. The parallel current associated with the twist will have a transverse scale a comparable to the size of the photospheric vortex of granular dimensions. The current density associated with current of this scale length is too small to provide the required Joule heating. Two mechanisms for reducing the transverse scale have been suggested. In a series of papers, Parker[9,10] has argued that a force-free magnetic configuration ($\boldsymbol{j} \times \boldsymbol{B} = 0$) continuously deformed at the base by photospheric motion will eventually generate a singularity in which the current that flows along the loop contracts into a sheet current of zero thickness, i.e. a tangential discontinuity (in the MHD limit). This conclusion has recently been bolstered by an optical analogy[11] to the equation of force free equilibria:

$$\boldsymbol{\nabla} \times \boldsymbol{B} = \alpha \boldsymbol{B} \ , \quad \boldsymbol{B} \cdot \boldsymbol{\nabla} \alpha = 0 \ , \quad \alpha = \boldsymbol{B} \cdot \boldsymbol{\nabla} \times \boldsymbol{B}/B^2 \ . \tag{2}$$

Once the current contracts into a singular sheet the ohmic heating rate increases by orders of magnitude and a significant fraction of the magnetic energy in excess of the vacuum field[12] could be released in thermal and energetic particles. However, Van Ballegooijen[13] has contended that long before the current contracts into a tangential discontinuity it filaments, and this filamentation contains all scales less than the scale a at the base of the flux tube. This follows from the following considerations. Let the current density $\boldsymbol{j} = \boldsymbol{j}_\| + \boldsymbol{j}_\perp$ where $\|$ and $\perp$ refer to parallel and perpendicular to the vacuum magnetic field $\boldsymbol{B}_o$. Because of the force free condition $\boldsymbol{j}_\perp = 0$ and since the divergence of $\boldsymbol{j}$ must vanish, we obtain:

$$\boldsymbol{\nabla} \cdot \boldsymbol{j}_\| = \frac{\partial}{\partial z} j_\| + \frac{1}{B_o} \boldsymbol{\nabla}_\perp A \times \hat{\boldsymbol{z}} \cdot \boldsymbol{\nabla}_\perp j_\| = 0 \ , \tag{3}$$

where $\boldsymbol{A} = A\hat{\boldsymbol{z}}$ is the vector potential of the field $\boldsymbol{B}_\perp$ generated by $j_\|$. Note that $j_\| = -\nabla_\perp^2 A$ and $\boldsymbol{B} = B_o \hat{\boldsymbol{z}} + \boldsymbol{B}_\perp$ where $\boldsymbol{B}_\perp = \boldsymbol{\nabla}_\perp A \times \hat{\boldsymbol{z}}$: Furthermore, $|\boldsymbol{B}_\perp| \ll B_o$ and the unit vector along $\boldsymbol{B}$ may be approximated by $\hat{\boldsymbol{\epsilon}}_\| = \hat{\boldsymbol{z}} + \boldsymbol{\nabla}_\perp A \times \hat{\boldsymbol{z}}/B_o$. Given the current density at $z = 0$, Eqn. (3) provides the current density at any z. Numerical solutions of Eqn. (3) by Longcope and Sudan[14] confirm Van Ballegooijen's analysis.

In a realistic situation the configuration of the injected current is continuously changing because of vortex motion at $z = 0$. Equation (3) must be supplemented by the statement that the electric field along the magnetic lines of force vanishes because of the high conductivity of the plasma. Expressed in terms of A and ϕ the vector and scalar potentials this condition becomes

$$\frac{\partial \phi}{\partial z} + \frac{1}{B_o} \boldsymbol{\nabla}_\perp A \times \hat{\boldsymbol{z}} \cdot \boldsymbol{\nabla}_\perp \phi = -\frac{\partial A}{\partial t} \ , \tag{4}$$

and the plasma velocity is derived from $v_\perp = -c\nabla_\perp \phi \times \hat{z}/B_o$, i.e. the $E \times B$ velocity. The velocity $v_\perp$ is prescribed at $z = 0$ by $\phi(x_\perp, t, z = 0)$. The pair of equations (3) and (4) represent a slowly deforming flux tube with prescribed ϕ and A at $z = 0$. These equations operate on the quasi-static time scale which is much longer than the Alfvén transit time L/v_A; in fact, there are no dynamical terms and time is treated as a parameter. During such a quasi-static evolution, the system eventually approaches a state of neutral stability which separates stable from unstable equilibria. At this state, there is a nonvanishing disturbance in the flux tube even without any motion[14] at $z = 0$ and $z = L$. The system drops from this state to the next available neighbouring stable state with lower potential energy. The time scale of this transition is determined by dynamical equations governing plasma motion. Thus at this stage, the quasi-static equations have to be abandoned because of the abrupt change in time scales.

EVOLUTION OF SPATIAL STRUCTURE

The appropriate equations to replace the quasi-static equations (3) and (4) are the magnetohydrodynamic equations. If the transverse dimension of the loop $a \ll L$ we can adopt a much simpler version viz., the so-called "reduced MHD" equations[5,15,16] that deal with the vorticity $\Omega = \hat{z} \cdot \nabla \times v_\perp$ and the vector potential A.

$$\frac{\partial \Omega}{\partial t} + v_\perp \cdot \nabla_\perp \Omega = \frac{\partial}{\partial z} J + B_\perp \cdot \nabla J + \nu \nabla_\perp^2 \Omega , \qquad (5)$$

$$\frac{\partial A}{\partial t} + v_\perp \cdot \nabla_\perp A = -\frac{\partial \phi}{\partial z} + \eta \nabla_\perp^2 A , \qquad (6)$$

with $\Omega = \nabla_\perp^2 \phi$, $J = -\nabla_\perp^2 A$; ν and η are the plasma viscosity and resistivity respectively. These equations are expressed in units of L/v_A(time), L (z coordinate) and $a/2\pi$ (x and y coordinates). These equations reduce to the quasi-static equations by setting $\nu = \eta = 0$ and the inertial terms $\partial \Omega/\partial t = 0 = v_\perp \cdot \nabla_\perp \Omega$. The photospheric drive is introduced by prescribing

$$\Omega(x_\perp, t, z = 0) = f^0(x_\perp, t), \quad \Omega(x_\perp, t, z = 1) = f^1(x_\perp, t) ,$$

to be randomly generated fields with Gaussian statistics and

$$\tau_E = 2\pi / <|f^j(x_\perp,t)|^2>^{\frac{1}{2}}; \quad \tau_c = \tau_E^2 \int_0^\infty d\tau < f^j(x_\perp,t) f^j(x_\perp, t+\tau) > .$$

This forcing is maintained quasi-static by demanding $\tau_E, \tau_c \ll 1$.

A three-dimensional code MARLO which adopts a spectral technique in x and y (because of periodic conditions) and a finite difference numerical scheme in z has been developed[14]. Studies on this code on a 64×64×10 grid have shown the evolution of equilibria of increasing spatial complexity. In particular, one can observe that the current filaments into structures of thinner and thinner section. This feature confirms the analysis of Van Ballegooijen[13] and simulations by Mikic et al.[17] of the full set of MHD equations. The simulations of Eqns. (5) and (6) take less computer time than the full MHD equations. The MARLO code takes 40µsec/timestep/fourier mode on the IBM 3090/600 machine at the Cornell National Supercomputing Facility. Nevertheless, the run time in such spectral codes is limited, in the absence of dissipation, by the appearence of structures at the smallest scales, i.e the grid size. If dissipation is included then a steady state can be reached but capturing the inertial range is difficult.

However, it can be shown analytically through a perturbation analysis[14] and confirmed numerically that the spectrum of the current density J_k where $k = (k_x, k_y)$ is initially exponential:

$$< |J_k|^2 > \sim \exp -\lambda(t)|k| \qquad (7)$$

where $\lambda(t)$ decreases to zero, at which point the perturbation scheme breaks down.

In fact, the representation of the normal modes of order 10^{15} of an actual solar loop by the finite modes of a numerical simulation must indeed overlook the fine structure. To make progress, the principle of scale invariance must be invoked to represent the large number of modes that cannot be accounted for in a typical numerical simulation. At each scale, the dissipative coefficients are functionals of the large scale modes computed on the grid. The fine scales driven by the large scales are not computed in detail but their action on the large scales is taken into account by the revised dissipative terms. A detailed calculation[18] based on a Renormalization Group analysis (RNG) of Eqns. (5) and (6) leads to the following expressions for the renormalized viscosity and resistivity in terms of the grid size Δx:

$$\nu_* = (1 + \rho_*)\alpha(\rho_*)\frac{(\Delta x)^2}{\pi^2}[(1 + \rho_*)\Omega^2 + (1 - \rho_*)J^2]$$

$$\eta_* = (1 - \rho_*)\alpha(\rho_*)\frac{(\Delta x)^2}{\pi^2}[(1 + \rho_*)\Omega^2 + (1 - \rho_*)J^2]$$

where

$$\rho_* = \left(1 + 4\frac{|\nabla_\perp \phi|^2}{|\nabla_\perp A|^2}\right)^{-\frac{1}{2}}$$

and $\alpha(\rho_*)$ is a function that ranges between values at 2 and 7; Ω, J and A are computed on the numerical grid. This RNG based subgrid modelling will be included in a revised version of the MARLO code.

IMPULSIVE EVENTS IN THE QUASI-STATIC EVOLUTION

We have noted previously that dissipationfree spectral codes have only a limited range in time before numerical inaccuracies caused by the development of fine scales of order grid size set in. In order to study the long time behavior of coronal magnetic loops, say over 10^6 Alfvén wave times, the only alternative is to limit the spatial accuracy of the code. The motions described by Eqns. (5) and (6) obey the energy relation

$$\frac{d}{dt}(E_M + E_k) = P_F - P_\eta - P_\nu \qquad (8)$$

where the magnetic energy $E_m = \frac{1}{2}\int d^3x\, |\nabla_\perp A|^2$, the kinetic energy $E_k = \frac{1}{2}\int d^3x\, |\nabla_\perp \phi|^2$, the ohmic dissipative power $P_\eta = \eta \int d^3x\, |J|^2$, the viscous dissipative power $P_\nu = \nu \int d^3x\, \Omega^2$ and the input power from photospheric motion is $P_F = \int d^2x_\perp \{A(x_\perp, z=1)f^1(x_\perp,t) - A(x_\perp, z=0)f^0(x_\perp,t)\}$. Recalling that the eddy turnover time τ_E and the eddy correlation time τ_c are of order $10^3 \sim 10^5$ sec for granular and supergranular motion respectively, the quasic-static limit corresponds to $\tau_E \gg 1$. If the underlying system is quasi-static then in a series of numerical runs simulating Eqns (5) and (6) with $\tau_E \gg 1$ we would expect:

$$J(x_\perp, z, t) \to \tilde{J}(x_\perp, z, t/\tau_E)$$

$$\Omega(x_\perp, z, t) \to \tau_E^{-1}\tilde{\Omega}(x_\perp, z, t/\tau_E)$$

$$E_M(t) \to \tilde{E}_M(t/\tau_E) \;;$$

$$E_k(t) \to \tau_E^{-2}\tilde{E}_k(t/\tau_E)$$

$$P_F(t) \to \tau_E^{-1}\tilde{P}_F(t/\tau_E); P_\nu(t) \to \tau_E^{-2}\tilde{P}_\nu(t/\tau_E) \;.$$

In steady state, it is necessary that P_ν and P_F balance each other; thus, we hypothesize a non-quasi-static component of $P_\nu(t)$ such that:

$$<P_\nu(t)> \sim \tau_E^{-1} \;. \qquad (9)$$

In order to test the above hypothesis a low-dimensional analog[19] of Eqns (5) and (6) has been constructed as follows. The perpendicular spatial dependence of all scalar fields in these equations are represented using only three orthogonal basis functions

$$\psi_1(x_\perp) = \sqrt{2}\cos(x),\; \psi_2(x_\perp) = \sqrt{2}\cos(\alpha y),\; \psi_3(x_\perp) = 2\sin(x)\sin(\alpha y) \;. \qquad (10)$$

α is a fixed parameter not equal to unity. Projecting Eqns (5) and (6) onto these basis functions yields a set of six coupled partial differential equations in z and t:

$$\partial\vec{\Omega}/\partial t + \vec{\phi}\times\vec{\Omega} = \partial\vec{J}/\partial z - \vec{A}\times\vec{J} - \nu L\cdot\vec{\Omega} \qquad (11)$$

$$\partial\vec{A}/\partial t + \vec{\phi}\times\vec{A} = -\partial\vec{\phi}/\partial z - \eta L\cdot\vec{A} \qquad (12)$$

$$\vec{\Omega} = -L \cdot \vec{\phi}, \quad \vec{J} = L \cdot \vec{A} \qquad (13)$$

Any scalar such as the current density J when expressed in terms of the basis functions is given by three coefficients that depend upon z and t which are combined into an iso-vector $\vec{J}(z,t)$. The linear operator L is a projection of $-\nabla_\perp^2$ and $L \equiv \text{diag}\{1, \alpha^2, 1 + \alpha^2\}$. The cross-products represent the advective terms. The current density and vorticity are represented on alternating points of uniform grid in z. The current density isovectors $\vec{J}^o, \vec{J}^1, \vec{J}^2, \vec{J}^3$ are defined at $z = 0, 1/3, 2/3,$ and 1 respectively while $\vec{\Omega}^1, \vec{\Omega}^2, \vec{\Omega}^3$ are located at $z = 1/6, 1/2$ and $5/6$. The vorticity is defined at $z = 0$ and $z = 1$ through f^o and f^1 the external drivers. This leads to a system of 21 O.D.E.'s which is the low dimensional analogue of the original P.D.E.'s in x, y, z and t. In the corona, the resistive decay time is very large so that we set $\eta = 0$ which allows an equilibrium to exist for ever if not externally perturbed. Oscillations about such an equilibrium leads to Alfvén waves whose damping depends upon ν the viscosity. To obtain a critical damping of Alfvén waves, we set $\nu = 2$. When the footpoints are twisted slowly with $\tau_E \gg 1$ the state of the system changes slowly passing from one equilibrium to the next until a marginally stable state is reached. When pushed beyond this state the magnetic energy drops abruptly and the kinetic energy and viscous damping power experience abrupt spikes lasting $\sim 3\tau$. Before and after this transition the kinetic energy is a factor of $\sim \tau_E^{-2}$ times the magnetic energy indicating the system is almost in equilibrium. Thus, the quasi-static evolution is punctuated by an impulsive event which corresponds to a "loss of equilibrium"[20,21].

Figure 1 shows the section of one long run[19] with $\nu = (2\pi)^2/20 \simeq 2, \tau_E = 300$ and $\tau_C = 100$. The magnetic energy fluctuates continuously in response to the randomly varying footpoint motion, sometimes dropping abruptly. The viscous damping power has slow quasi-static fluctuations at levels around $\nu \tau_E^{-2} \sim 10^{-5}$, and short, strong relaxation spikes with amplitudes as high as 10. These two components can be distinguished using a viscous power threshold; in this case we used $P_\nu = 10^{-3}$. A single relaxation event is defined to begin when P_ν first exceeds this threshold and to end when it falls below the threshold for more than one Alfvén time.

The drop in magnetic energy, ΔE_M, from a relaxation event can be estimated by integrating the viscous power $\Delta E_M = \int dt P_\nu(t)$. By choosing the viscosity so that relaxation events are as short as possible (few last longer that four Alfvén times) we assure that the footpoints can not exchange a significant portion of the energy during a relaxation.

During a full run of 10^6 Alfvén times, of which Fig. 1 is a section, there were 1093 such events. Figure 2 shows the distribution of amplitudes, ΔE_M, for these events. It shows that larger amplitude events are less frequent, decreasing at a slope close to ΔE_M^{-1}. The lower cutoff of this distribution is due to the threshold used for detecting events. The amplitudes of events which peak just above the threshold are likely to be underestimate, leading to inaccuracies on the left of the plot.

Also shown in Fig. 2 is the distribution of intervals between successive events, Δt. These are exponentially distributed with a mean value of 915 Alfvén times or about $3\tau_E$. In other runs with different values of τ_c and τ_E the mean

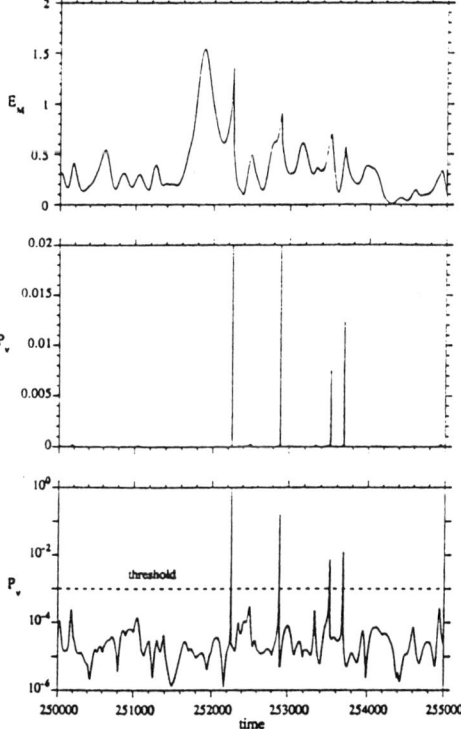

Fig. 1. Section of the time history of a long run with multiple impulsive events, showing magnetic energy, E_M and both linear and logarithmic plots of the viscous dissipation rate, P_ν.

interval remains about $3\tau_E$, independent of the correlation time τ_c. Exponential distributions are often associated with Poisson processes, that is processes where the probability of an event does not depend on the time since the previous event. Since this system is driven by noise it is not surprising that it has such a property.

Finally, it is worth noting that the amplitude of an event is not correlated with the time since the previous event. Figure 2c shows a scatter of plot of Δt against ΔE_M for all 1093 events with no apparent correlation. This would not be the case if the magnetic field built up energy continuously until it was released by a loss of equilibrium. Instead, the magnetic field both gains and loses energy randomly to the footpoints during the interval between events. This is possible because there are enough degrees of freedom for the system to change its relation to the footpoints.

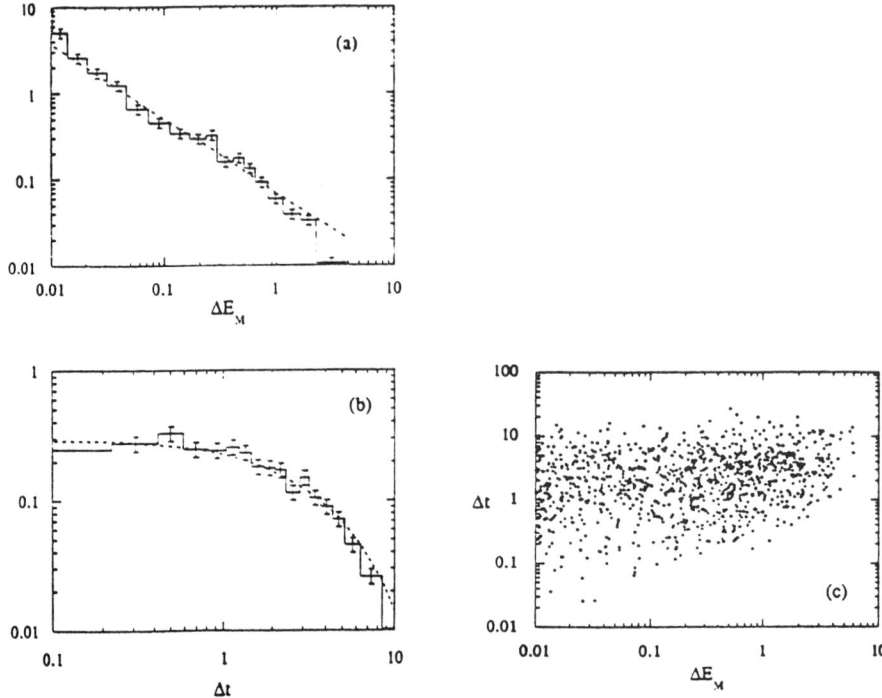

Fig. 2. (a) Histogram of event amplitudes, ΔE_M; dashed line is $0.068 \Delta E_M^{-0.87}$. (b) Histogram of time interval between successive events, $\Delta t / \tau_E$; dashed line is $0.3 \exp(-\Delta/3.3\tau_E)$. (c) Scatter plot of $\Delta t / \tau_E$ vs. ΔE_M.

CONCLUSIONS

From the above discussion, we conclude that the heating of the coronal magnetic loops is a direct consequence of photospheric motion. The fast motions can trigger shear Alfvén waves directly which are absorbed in a few bounces by damping on the regions of stochastic field lines in these magnetic configurations. The slower motions twist and untwist the magnetic field quasi-statically. The current filaments into fine structures which enhance the heating rate but it is not clear how long it would take to reach scale sizes for which Joule heating would become significant. The quasi-static motion is punctuated by impulsive events or nanoflares during which the kinetic energy increases abruptly giving rise to Alfvén waves which are again damped on stochastic field lines.

The fraction of energy dissipated in these impulsive events is a^2/L^2 of the vacuum field energy in the photospheric motion. The rate at which this energy is dissipated per unit area of the photosphere is

$$P_d = (a^2/L)(B_o^2/8\pi)\tau_E^{-1}$$

(τ_E is expressed in seconds) assuming that all the energy released in an impulse event is absorbed in the loop. This assumption needs to be checked by computing the fraction of the twisted flux tube volume that contains stochastic field lines. The dependence of τ_E on the observed scale of photospheric motions is required to be evaluated P_d and thereby complete this theoretical picture.

Finally, towards the completion of this work our attention was directed to early work by Gold[22,23]. The qualitative picture of the abrupt relaxation of the magnetic energy of coronal magnetic loops caused by footpoint motion postulated by Gold, receives quantitative verification in our analysis.

ACKNOWLEDGEMENTS

This work was supported by NSF Grant #90-20719 and computations performed at NSF/IBM supported Cornell National Super Computing Center.

REFERENCES

1. G.L. Withbroe and R.W. Noyes, Ann. Rev. Astron. Ap. 15, 363 (1977).
2. See for example: J. Heyvaerts and E.R. Priest, Astron. Astrophys. 117, 220 (1983).
3. J.V. Hollweg, Ap. J. 277, 392 (1984).
4. P.L. Similon and R.N. Sudan, Ap. J. 336, 442 (1989).
5. R.N. Sudan and P.L. Similon, Proceedings of the Joint Varenna-Abastumani International School and Workshop on Plasma Astrophysics, Varena, Italy, (ESA SP-285) (1988).
6. R.N. Sudan, in Mechanisms of Chromospheric and Coronal Heating, P. Ulmschneider, E.R. Priest, R. Rosner (eds.), (Springer-Verlag, 1991), p. 448.
7. P.L. Similon and S. Zargham, in Mechanisms of Chromospheric and Coronal Heating, P. Ulmschneider, E.R. Priest, R. Rosner (eds.), (Springer-Verlag, 1991), p. 438.
8. E.N. Parker, Physics Today, july 1987, 36 (1987).
9. E.N. Parker, Ap. J. 264, 635 (1983a).
10. E.N. Parker, Ap. J. 264, 642 (1983b).
11. E.N. Parker, Phys. Fluids B 3, 2652 (1991).
12. J.J. Aly, Ap. J. 283, 349 (1984).
13. A.A. Van Ballegooijen, Ap. J. 331, 1001 (1986).
14. D.W. Longcope and R.N. Sudan, to appear in Ap. J., Jan. 1992.
15. H.R. Strauss, Ap. J. 326, 418 (1988).
16. M.N. Rosenbluth, D.A. Monticello, H.R. Strauss and R.B. White, Phys. Fluids 19, 1987 (1976).
17. Z. Mikic, D.D. Schnack and G. Van Hoven, Ap. J. 338, 1148 (1989).
18. D.W. Longcope and R.N. Sudan, Phys. Fluids B 3, 1945 (1991).
19. D.W. Longcope and R.N. Sudan, Impulsive Events in the Quasi-Static Evolution of a Forced Non-Linear System, LPS Report 91-12, Laboratory

of Plasma Studies, Cornell University, Ithaca, NY (1991).
20. J.A. Klimchuck and P.A. Sturrock, Ap. J. 345, 1034 (1989).
21. B.C. Low, Ap. J. 212, 234 (1977).
22. T. Gold, Proc. AAS-NASA Symp. on Physics of Solar Flares, (W.N. Hess (ed.), NASA Washington D.C., 1964) p. 389.
23. T. Gold, I.A.U. Symp. 22, (R. Lüst (ed.), North-Holland Co., Amsterdam, 1963) p. 390.

CORONAL HEATING THROUGH LACK OF MHD EQUILIBRIUM

P.C.H. Martens
Lockheed Palo Alto Research Laboratories, CA 94304-1191, USA

M.T. Sun and S.T. Wu
CSPAR/University of Alabama in Huntsville, AL 35899, USA

ABSTRACT

We present an analytical example of a series of magnetostatic equilibria with an endpoint. Numerical simulation demonstrates that oscillatory behaviour sets in at the endpoint, with a typical amplitude of 50 km/sec. We suggest this in situ wave generation is an energy source for coronal heating.

INTRODUCTION

In the contribution by Sudan[1] there was a demonstration of "loss (or, better said, lack) of equilibrium" in the reduced, incompressible, MHD equations. Here we will demonstrate a similar result in the set of full, compressible, MHD equations, and therefrom deduce a scenario for the origin of coronal heating.

Parker, in a long series of papers spanning almost two decades (e.g. refs. 2,3), has claimed that the coronal magnetic field, even when evolving in response to smooth continuous photospheric footpoint motions, will soon be unable to achieve a smooth, force-free equilibrium, and develop tangential discontinuities. This is thought to lead to enhanced dissipation in current sheets, perhaps spiky in nature, and provide the energy release responsible for coronal X-ray emission.

The results presented here indicate that "loss of equilibrium" can indeed occur, but we do not find the development of tangential discontinuities, or any significant concentration of current. Instead magnetic energy is dynamically released in the form of wave motions. This leads us to a new scenario for coronal heating, combining aspects of the so-called AC and DC scenarios (see Gómez[4] for a recent review).

We suggest that resonant absorption of MHD waves, as originally proposed by Ionson[5], provides the source of coronal heating. However, the waves do not enter the corona from the underlying photosphere/chromosphere, but are generated in situ, deriving their energy from the DC coronal currents after the onset of non-equilibrium.

In the following section we will derive an analytical sequence of MHD equilibria evolving towards an endpoint, at which no further analytical equilibrium exists. Then we will numerically analyze the evolution as the endpoint is approached, and compare our numerical results with those of others and against observations of coronal line-broadening.

A BOUNDED SERIES OF FORCE-FREE EQUILIBRIA

We consider a force-free magnetic arcade straddling a photospheric neutral line. The arcade has translational symmetry along the neutral line, and rotational symmetry about an axis below the surface. Our Cartesian coordinate system has z denoting the height above the photosphere, x the projected distance from the neutral line, and y the dummy coordinate along the neutral line. A parameter t denotes the depth of the symmetry axis below the photosphere.

It is convenient to work in a cylindrical coordinate system (r, φ, y) centered at the symmetry axis $z = -t$. The radial coordinate in this system $r^2 = (z+t)^2 + x^2$, and the partial derivatives ∂_y and ∂_φ vanish by definition. In this system, the force-free equation reduces to

$$\frac{d}{dr}(B_\varphi^2 + B_y^2) + \frac{2B_\varphi^2}{r} = 0, \tag{1}$$

with $B_r = 0$ everywhere, because the fieldlines are divergenceless.

We impose the following normal component of the magnetic flux at the photosphere:

$$B_z(z = 0, t) = x \exp(-x^2). \tag{2}$$

Note that the total photospheric flux on each side of the neutral line is finite, and that the normal field is independent of the parameter t – hence only shearing photospheric motions are allowed.

The solution of Eq. (1) for the boundary condition Eq. (2) is

$$B_x = -(z+t)B_0 \exp[-(x^2 + z^2 + 2zt)/2] \tag{3}$$

$$B_y = B_0\{[1 - x^2 - (z+t)^2]\exp(-x^2 - z^2 - 2zt) + C^2\}^{1/2} \tag{4}$$

$$B_z = xB_0 \exp[-(x^2 + z^2 + 2zt)/2], \tag{5}$$

with B_0 an arbitrary scaling factor, and the parameter C^2 representing the axial field at infinity ($C^2 = B_y(\infty)^2/B_0^2$). We take this field to represent the overlying field of the active region of which the arcade under consideration is part.

The shear displacement along the neutral line of the photospheric footpoints of the fieldlines is given by

$$\Delta y(z = 0, t) = 2 \arctan(x/t)\sqrt{1 - t^2 - x^2 + C^2 \exp(x^2)}. \tag{6}$$

This shear displacement is continuous at $x = 0$, except at $t = 0$, a value which we henceforth exclude from consideration.

For a physically meaningful solution the factor under the square root in Eqs. (4) and (6) has to be nonnegative. This leads to the constraints

$$t^2 \leq C^2 + 1 \quad for \quad C^2 \geq 1 \tag{7}$$

$$t^2 \leq 2 + 2\ln(C) \; for \; C^2 \leq 1 \tag{8}$$

$$C^2 \geq 1/e^2. \tag{9}$$

Obviously our parameter t is intended to denote time, and the above solutions to represent a time sequence of quasistatic equilibria. We note that the quasistatic evolution can made to be as slow as one pleases by replacing t with εt. In the next section we will investigate is what happens as t approaches and surpasses its upper limit. We emphasize that the sequence defined above constitutes a valid thought experiment in the sense that a footpoint displacement which is everywhere finite and continuous – and therefore physically possible – drives the evolution.

A NUMERICAL EXPERIMENT

A numerical code for solving the time dependent MHD equations using a new Nimble Implicit Continuous Eulerian (NICE) integration scheme has been developed by M.T. Sun[6]. The code is very well suited to simulate solar-type MHD problems with a combination of reflecting and non-reflecting boundaries, compressibility, and large variations in plasma β. The code has successfully solved a number of test problems. Details are given in Sun[6].

In our numerical analysis we have used the ideal, compressible, MHD equations with zero fluid viscosity and zero gravity. (The latter will be justified by considering an arcade with height considerably smaller than the coronal temperature sale height). The ideal gas law is used, and the energy equation is adiabatic. To save computing time we have also imposed symmetry along the neutral line, $\partial_y = 0$, but not cylindrical symmetry.

In the simulation described here we used $C = 0.4$, start at $t = 0.9 \times t_{max}$, and continue until approximately $1.3 \times t_{max}$. The unit for the magnetic field is 45 Gauss, the plasma β just above the neutral line is 0.1, and the plasma temperature $T = 3 \times 10^6 K$. This leads to realistic values for coronal pressure and density. We chose our length unit such that the physical size of the computational domain is 28,000 km in the x direction, and 12,500 km in the z direction (much smaller than the temperature scale-height). The grid is 36 $\times$ 25. The shear velocity scale is 1 km/sec leading to a maximum shear velocity of 6 km/sec at one time at one point (later simulations achieved lower maximum velocities). The time unit is 5000 sec.

The results of the simulation are depicted in Figure 1. The sequence of panels with projected fieldlines shows clearly that up to about t_{max} the numerical results closely follows the analytical equilibrium sequence. After that the fieldlines start to oscillate reaching an amplitude of about 50 km/sec. This result clearly suggests that "loss of equilibrium" has occurred. Since we are using the ideal MHD equations no reconnection can take place, and an eruption is not to be expected with the infinite amount of overlying flux.

This is not unlike the real situation for a small arcade of loops embedded in a larger active region. With a coronal magnetic Reynolds number of about 10^{12}, and with the numerical results giving no indication of strong current concentrations, indeed no significant reconnection can occur on a time-scale of up to 100 Alfvén crossing times (the duration of the simulation). Therefore the conversion of magnetic energy into MHD wave energy, may be the only efficient method to shed excess free energy.

A NEW SCENARIO FOR CORONAL HEATING

We cannot conclude at this point that "loss of equilibrium" has been unambiguously demonstrated, since it is possible that linear stability is simply lost as time approaches t_{max}. However, work by Cargill et al.[7] on the stability of force-free line-tied coronal arcades has not shown instability for any of a number of cases quite similar to ours. Cargill[8] concludes "The bottom line is that there is strong evidence that force-free arcades are absolutely stable ..". Analysis of the linear stability of the equilibrium sequence of Sect. 2 is in progress.

We note that for our scenario of coronal heating, summarized in the intro-

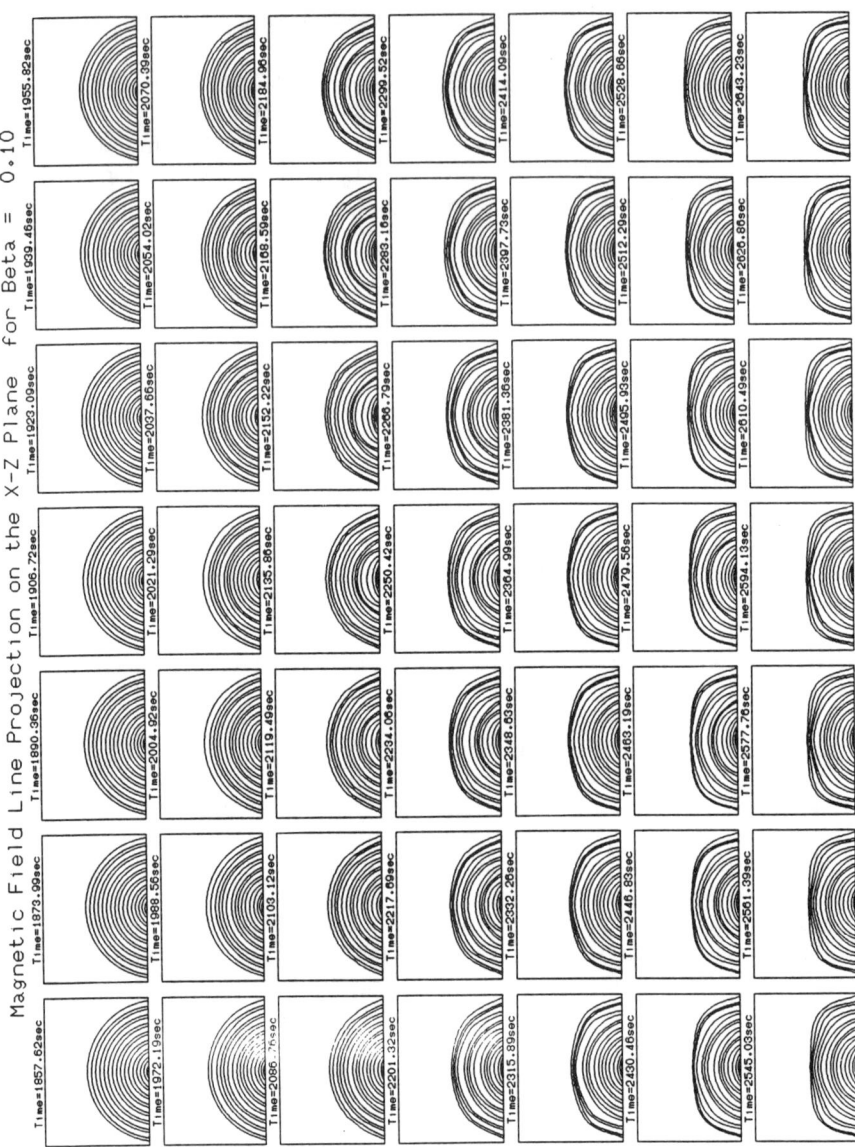

Fig. 1. Numerical simulation of the evolution of the magnetic field in projection on the $x - z$ plane.

duction, it makes no difference whether the dynamic behaviour originates from loss of stability or of equilibrium. We find that free magnetic energy is build up in the corona as a result of slow photospheric footpoint motions driving the coronal field through a sequence of force-free equilibria of increasing energy. Then, as either non-equilibrium or instability sets in, part of the magnetic energy is rapidly released in the form of MHD-waves. Resonant absorption of these waves may heat the coronal plasma to X-ray emitting temperatures.

The in situ generation of the waves removes the need for a large wave-energy flux (10^{6-7} erg cm^{-2} sec^{-1}) entering the corona to compensate for radiative losses – a major weakness in most wave-heating theories since observations from OSO VIII (Athay and White[9]) have put a much lower upper limit on this flux.

In addition, the wave amplitude of about 50 km/sec we found is consistent, both in magnitude and location, with the non-thermal line broadening observed over neutral lines in active regions by the Flat Crystal Spectrometer (FCS) aboard SMM (Saba and Strong[10]).

We suggest that frequent occurrence of the "loss of equilibrium" and wave generation reported here can release enough energy to maintain the X-ray corona. If the energy release occurs on intervals larger than the cooling time of the plasma, the heating becomes "episodic" and the observed form of the coronal differential emission measure may be recovered as well (see Sturrock et al.[11]).

REFERENCES

1. R.J. Sudan, these proceedings , (1991).
2. E.N. Parker, Ap. J. 174, 499 (1972).
3. E.N. Parker, Ap. J. 318, 876 (1987).
4. D.O. Gómez, Fund. of Cosmic Phys. 14, 131 (1990).
5. J.A. Ionson, Ap. J. 226, 650 (1978).
6. M.T. Sun, Three-Dimensional Time-Dependent Fluid Flow Simulations: Ordinary and Magnetohydrodynamic (MHD) Fluids (Dissertation, The University of Alabama in Huntsville, 1991).
7. P.J. Cargill, A.W. Hood, and S. Migliuolo, Ap. J. 309, 402 (1986).
8. P.J. Cargill, private communication , (1991).
9. R.G. Athay and D.R. White, Ap. J. 226, 1135 (1978).
10. J.L.R. Saba and K. Strong, Ap. J. 375, 789 (1991).
11. P.A. Sturrock, W.W. Dixon, J.A. Klimchuk, and S.K. Antiochos, Ap. J. Letters 356, L31 (1990).

ON THE COLLECTIVE APPEARANCE OF CORONAL LOOPS AND THE RESISTIVE HEATING INSTABILITY

Yu-Qing Lou
Geophysical Institute
University of Alaska Fairbanks, Fairbanks, AK 99775-0800, USA

ABSTRACT

We investigate the onset conditions for direct resistive heating instabilities coupled with radiative processes within a twisted magnetic flux rope of axisymmetry and relate the helical patterns wrapping along the rope initiated by such instabilities to the collective appearance of compact X-ray loops in a certain phase of active region development in the solar corona. Since the emergence and the subsequent evolution of a gigantic magnetic flux rope in the solar atmosphere involve complicated physical processes, it is expected that such instabilities occurring in an ensemble of many current sheaths embedded in a stressed, twisted, and bulged magnetic flux rope will manifest as collective X-ray loop structures on various spatial scales and with varieties of large-scale morphologies.

INTRODUCTION

X-ray images of the Sun reveal extremely inhomogeneous structures in its corona[1,2] and clearly indicate the association of magnetic fields in forming various coronal structures.[3,4] Notable among coronal structures is the ubiquitous appearance of compact X-ray loops in active regions. In general, three phases may be recognizable in the evolution of a typical active region, viz., the initial emerging stage, the lasting stage of compact X-ray loops, and the final diffused stage.[5,6] By carefully examining X-ray pictures of the Sun, one can usually identify a few examples of active regions in which a collection of fairly regular loop patterns seems to betray an underlying gigantic magnetic flux rope. Based on such a scenario, we suppose that the concurrence of regular X-ray loop patterns around a flux rope is very likely associated with some sort of instability. And this motivates our recent investigation[20] on resistive heating instabilities coupled with radiative processes. Thus our perspective, complementary to those of earlier studies,[7,8,9,10] for the structure of compact X-ray loops is emphatically collective rather than individual.

The basic concept of resistive heating instabilities fits into the general category of thermal instabilities.[11,12] However, since the generalized heat-loss function in formulations of thermal instabilities usually does not include the current dissipation[12] and also because the significance of the concept of resistive heating instability has been realized in several important applications, it seems appropriate to distinguish the class of resistive heating instabilities from the broad class of thermal instabilities.[13] Briefly speaking, resistive heating instabilities are caused by the temperature dependence of electrical conductivity. Considerable complication for such instabilities arises in the presence of various magnetic field geometries. In general, constant phase line for such instabilities tends to align with the background magnetic field line. We note here that several terminologies have been adopted in the conceptual development of resistive heating instabilities. For example, "superheating instability",[14,15] "Joule mode instability",[16,17] "current filamentation instability",[18,24] and "resistive-heating instability"[19,20] all refer to a similar physical process.

A MODEL FORMULATION

Our model formulation for compact X-ray loop structures involves several macroscopic processes such as current dissipation, cross-field thermal conduction, and radiative loss. It is known[5] that classical description of these processes given the coronal conditions would imply an extremely small spatial scale for current dissipation, which is difficult to be related to the observed scales of $10^8 \sim 10^9$ cm. On the other hand, it is also known from fusion experiments that when a driven current exceeds a critical level, the effective (anomalous) resistivity can increase by $3 \sim 5$ orders of magnitude due to collective wave-particle interactions.[21,22,23] Magnetized plasmas in the solar corona are likely in a turbulent state.[5] We thus presume that macroscopic transport processes in the presence of microscopic turbulence retain their classical forms but with increased effective magnitudes. Admittedly, this is a simplistic approach due to the lack of complete understanding for anomalous transport processes. To be specific, magnetic diffusivity η and thermal diffusion coefficient K are assumed in the classical forms $\eta^* T^{-3/2}$ and $\kappa^* T^{5/2}$, respectively, with η^* and κ^* 5 orders of magnitude larger than the corresponding classical values; the radiative loss term[5] is given by $n^2 \Phi(T)$ where n is the electron density and $\Phi(T)$ is a computed function of temperature T which may be approximated[7] by $5.4 \times 10^{-17} T^{-2/3}$ c.g.s.

We consider, as the background, a cylindrical electric current sheath embedded in a coaxial magnetic flux rope of axisymmetry. In the magnetohydrodynamic (MHD) formulation, this background is in inductive, thermal, and mechanical equilibrium. In other words, a helical magnetic field $\boldsymbol{B}$ consistent with a steady current in z-direction is described by $[0, B_\phi(r), B_z]$ with a constant B_z and a B_ϕ satisfying the static induction equation; the steady heating due to current dissipation is balanced by the cross-field thermal conduction and the radiative loss;[7] the radial Lorentz force associated with $B_\phi(r)$ is balanced by the pressure force; gravity is ignored for the sake of simplicity; and the ideal gas law is adopted. By solving a fourth-order system of equations, it is straightforward to obtain background variables in terms of r given prescribed parameters. We note that the mechanical equilibrium is not force-free but the low plasma beta condition can be achieved (as in the solar corona) for strong B_z.

The MHD perturbation equations can be readily obtained with respect to a specified background equilibrium. Since the background is axisymmetric, perturbations are characterized by $\exp(\sigma t + ikz + im\phi)$ dependence where σ is complex, k is a real wavenumber in z-direction, and m is an integer. In general, four major classes of instabilities may occur, viz., the class of MHD mechanical instabilities, the class of resistive tearing instabilities, the class of MHD thermal instabilities, and the class of resistive heating instabilities. The nature of the first one is the tendency of releasing magnetic stress via specific MHD modes of motion; that of the second is the tendency of accessing lower energy states by breaking field topology; that of the third is the tendency of releasing available thermal energy via specific modes of thermal-radiative exchange which is complicated by the presence of magnetic field; and that of the last is the tendency of releasing magnetic stress via current dissipation coupled with temperature variation of electrical conductivity.

Several MHD modes of oscillations can occur in the described background equilibrium such as modified versions of Alfvén, fast, slow, and surface modes etc. As the background current intensity is gradually increased, the manner that an instability sets in may be either direct (i.e., non-oscillatory) or oscillatory. We limit our consideration to the onset of direct instabilities for the moment and thus set $\sigma = 0$ in the MHD perturbation equations. It is noted that at the onset of direct instabilities, dynamic and energetic effects can be studied separately. Furthermore, for the onset of (direct) resistive heating instabilities, perturbations in magnetic field b and gas pressure p can be set to zero. From the r-component of the induction equation, mass conservation, ideal gas law, and the energy equation, we obtain the following equation in terms of temperature perturbation T, viz.,

$$\kappa^* T_o^2 \frac{d^2 T}{dr^2} + \left[5\kappa^* T_o \left(\frac{dT_o}{dr} \right) + \frac{\kappa^* T_o^2}{r} \right] \frac{dT}{dr}$$
$$- \left[\frac{5}{2} \kappa^* \left(\frac{dT_o}{dr} \right)^2 + \frac{A^2}{\pi \eta^*} - \left(\frac{31}{6} \right) T_o^{-3/2} \frac{p_o^2 \Phi(T_o)}{4 k_B^2 T_o^2} + \left(k^2 + \frac{m^2}{r^2} \right) \kappa^* T_o^2 \right] T \quad (1)$$
$$- \frac{3}{2} \eta^* T_o^{-3} \frac{m}{r^2} \frac{d(rB_\phi)}{dr} \left(\frac{mB_\phi}{r} + kB_z \right)^{-1} \left(\frac{p_o}{\rho_o} \frac{d\rho_o}{dr} - C_v \rho_o \frac{dT_o}{dr} \right) T = 0,$$

where constant A is proportional to the applied electric field, k_B is the Boltzmann constant, and C_v is the specific heat at constant volume. It is noted that the last term of (1) involves fluid compressibility, background inhomogeneity, and the temperature variation of electrical conductivity. The coefficient of T in the last term can be positive and sufficiently large to initiate the onset of (direct) resistive heating instabilities.

RESULTS AND DISCUSSION

The major characteristics within an equilibrium current sheath are the following. Pressure p_o and temperature T_o decrease with r; density ρ_o and B_ϕ increase with r. For the given conditions, the coefficient of T in the second term of (1) is usually negative. Thus, resistive heating instability cannot set in for axisymmetric (i.e., $m = 0$) perturbations. Whereas for $m < 0$, $k > 0$, and positive $mB_\phi/r + kB_z$ of small value (where $B_\phi > 0$ and $B_z > 0$ are assumed with no loss of generality), resistive heating instability sets in according to the described characteristics of the background. Furthermore, density perturbation is proportional to the negative of temperature perturbation due to the isobaric condition such that the perturbed radiative loss results in regularly spaced maxima and minima along the flux rope. In other words, the helical pattern initiated by the instability tends to align with the magnetic field line which supports the interpretation[5] that compact X-ray loops map out field lines. It is also clear from the last term of (1) that strong B_z tends to suppress a given instability whereas high current intensity promotes instability. For $|m| \gg 1$ and/or $|k| \gg 1$, the instability is suppressed due to the second term in (1). We have obtained critical solutions for various sets of parameters. It is noted that the radial node number of eigensolution T increases successively as the current intensity increases.

Since the estimated thickness of a current sheath is $\sim 10^5$ cm, thousands of current sheaths are needed to account for the X-ray brightness ($\sim 10^7$ erg cm^{-2} s^{-1}) of a typical active region. Unless one can follow the evolution of an active region with high temporal resolution, a typical X-ray picture probably reveals a stage where instabilities have fully developed. Thus, our identification of compact loop structures with thermal helical patterns initiated by resistive heating instabilities rests on an assumption that nonlinear evolution of such instability does not smear out completely linear spatial signatures. In summary, we suggest that resistive heating instabilities within an ensemble of many current sheaths embedded in a magnetic flux rope can be an important elemental mechanism leading to the pattern formation of compact X-ray loops. We further anticipate the evolution of an active region to be characterized by the competing processes such as resistive heating, mechanical, thermal, and resistive tearing instabilities.

It is a pleasure to acknowledge useful discussions with R. Rosner. This work was supported by the State Funding of Alaska and the NSF grant ATM-9014888.

REFERENCES

1. G.L. Withbroe and R.W. Noyes, Ann. Rev. Astron. Astrophys. 15, 363 (1977).
2. G.S. Vaiana and R. Rosner, Ann. Rev. Astron. Astrophys. 16, 393 (1978).
3. A.S. Krieger, G.S. Vaiana and L.P. Van Speybroeck, in Solar Magnetic Fields (Reidel, Dordrecht, 1971), p. 397.
4. P.S. McIntosh, A.S. Krieger, J.T. Nolte and G.S. Vaiana, Solar Phys. 49, 57 (1976).
5. R. Rosner, L. Golub, B. Coppi and G.S. Vaiana, Ap. J. 222, 317 (1978).
6. L. Golub, M. Herant, K. Kalata, I. Lovas, G. Nystrom, F. Pardo, E. Spiller and J. Wilczynski, Nature 344, 842 (1990).
7. R. Rosner, W.H. Tucker and G.S. Vaiana, Ap. J. 220, 643 (1978).
8. I.J.D. Craig, A.N. McClymont and J.H. Underwood, Astron. Ap. J. 70, 1 (1978).
9. A.W. Hood and E.R. Priest, Astron. Ap. J. 77, 233 (1979).
10. S.K. Antiochos and G. Noci, Ap. J. 301, 440 (1986).
11. E.N. Parker, Ap. J. 117, 431 (1953).
12. G.B. Field, Ap. J. 142, 531 (1965).
13. L. Sparks and G. Van Hoven, Ap. J. 333, 953 (1988).
14. B.B. Kadomtsev, in Review of Plasma Physics vol. 2 (Consultant Bureau, New York, 1966), p. 153.
15. D. Spicer, Solar Phys. 53, 305 (1977).
16. J. Heyvaerts, Astron. Ap. J. 37, 65 (1974).
17. A. Ferrari, R. Rosner and G.S. Vaiana, Ap. J. 263, 944 (1982).
18. D. Biskamp and W. Horton, Phys. Rev. Letters 35, 39 (1975).
19. L. Sparks and G. Van Hoven, Phys. Fluids 30, 2470 (1987).
20. Y.Q. Lou, Ap. J. , submitted (1991).
21. A.A. Galeev and R. Sagdeev, in Review of Plasma Physics vol. 7 (Consultant

Bureau, New York, 1979), p. 1.
22. D. Spicer, Space Sci. Rev. 31, 351 (1982).
23. E.D. Volkov, N.F. Perepelkin, V.A. Suprunenko and E.A. Sukhomlin, Collective Phenomena in Current-Carrying Plasmas (Gordon and Breach, New York, 1985).
24. G. Bodo, A. Ferrari, S. Massaglia, R. Rosner and G.S. Vaiana, Ap. J. 291, 798 (1985).

ALFVÉN WAVES IN CURRENT-CARRYING INHOMOGENEOUS PLASMAS

H. Shigueoka
UFF, Instituto de Física, 24024, Niterói, RJ, Brazil

C.A. de Azevedo
UERJ, Instituto de Física, 20550, Rio de Janeiro, RJ, Brazil

A.S. de Assis
UFF, Instituto de Matemática, 24024, Niterói, RJ, Brazil

P.H. Sakanaka
UNICAMP, Instituto de Física, 13081, Campinas, SP, Brazil

ABSTRACT

The Alfvén modes in inhomogeneous cylindrical current–carrying plasmas is studied solving numerically the Hain-Lüst equation. We have obtained the eigenfrequencies for different current density and it is shown that the distance of the frequencies from the lower edge of the Alfvén continuum depends on its profile. We have shown, using the WKB aproximation, that there exist a MHD discrete Alfvén mode. The numerical results of the discrete Alfvén frequencies are compared with the one obtained using the WKB approach. This study is suitable to understand oscillations and heating in solar prominences and also chromospheric–coronal heating and oscillations by Alfvén waves.

INTRODUCTION

To understand better the physics of the coronal heating and oscillation, prominence heating and oscillations, solar flare energy emission and other magnetized solar plasma events (see Heidelberg Conference 1990 and references therein)[1], the MHD theory can be a good tool as well as their main mode; the Alfvén wave. Therefore, we want to contribute to understand better these problems presenting some novel important features of Alfvén waves in plasmas with equilibrium current. The currentelless problem has been rather well studied (see Heidelberg Conference 1990 and references therein).

The spectrum of the ideal magnetohydrodynamics for cylindrical plasma systems with diffuse profile has two continua, known as the SLOW WAVE and ALFVÉN WAVE continua, due to the singularities of the eigenvalue equation. There are also two non-Sturmian regions, due to the zeroes of the coefficient of the highest order term, intercalated by two continua. Discrete eigenvalues exist in between these four regions, in particular, the discrete Alfvén modes are those found below the Alfvén continuum[2].

Experimental[3] and theoretical results[4,5] have shown the existence of stable discrete Alfvén waves (DAW) with frequencies below the lower edge of the Alfvén continuum for a given axial mode number n and azimutal mode number m, i. e., $\omega_{DAW}^2 < min[\omega_A^2(r)]$. We define $\omega_A^2 = (m+nq)^2 v_A^2/(qR)^2$, where q is the safety factor, R is the major radius, and v_A is the Alfvén speed. These modes are like stable kink modes and appear only under certain well-defined conditions. The study of the spectrum of a inhomogeneous plasma has become a topic of great interest because of the use of the Alfvén wave for plasma heating in controlled fusion devices, and because of the importance role to study oscillations in the solar loop prominences[6].

HAIN LÜST EQUATION

The discrete Alfvén waves appear on the magnetohydrodynamic spectrum of the diffuse linear pinches. On its stable side, the spectrum of the ideal magnetohydrodynamics for cylindrical plasma systems with diffuse profile has two continua: SLOW WAVE and ALFVÉN WAVE[2]. This arises as a result of the spread of two singularities, due to the diffuse profile in the eigenvalue differential equation. Besides these continua there are two regions where the coefficient of the second order term in the differential equation presents a pole. These are known as non-Sturmian regions. In between these four regions (two continua and two non-Sturmian), there are discrete eigenvalues[2]. These discrete modes accumulate toward a continuum as the radial mode number increases.

The discrete Alfvén frequencies with an arbitrary equilibrium magnetic shear are calculated numerically. The equilibrium field $\vec{B}$ is given by $B_r(r) = 0$, $B_\theta(r)$, $B_Z(r)$. From the first order perturbation of the MHD equations we obtain a second order differential equation on ξ_r, r-component of the plasma displacement by taking all perturbed quantities developed as $f(r)\ exp(-i\omega t + ikz + im\theta)$[7]. This equation is:

$$\frac{d}{dr}\left[f(r)\frac{d}{dr}\left(r\xi_r\right)\right] + g(r)\left(r\xi_r\right) = 0, \qquad (1)$$

where all the coeficients of this equation are expressed in Azevedo et al. (1991)[6].

The condition $f(r) = 0$ defines two continua, the shear Alfvén and the slow one. The shear Alfvén wave frequency is given by $\omega_A^2 = F^2/\rho$. This value is minimum at or near the plasma center and maximum at the plasma edge, $r = a$. The function $\omega^2 = \gamma p \omega_A^2/(\gamma p + B^2)$ defines the slow wave continuum.

ANALYTICAL STUDY OF THE DISCRETE EIGENMODES

At that point, we study the discrete Alfvén modes using an WKB analysis. Therefore, rather than introducing the most complete eigenvalue equation, here we adopt a simple model yet the essence of physics can be brought out. We will show that, within the framework of the MHD theory, when a bounded plasma is assumed, there will be discrete modes with the angular frequency ω below the shear Alfvén frequency. The spectrum of these modes depends on the plasma current, as a first order quantity.

We consider that the current density is relatively small so that $(B_\theta/B_z)^2 \sim \epsilon \ll 1$, that is, the twist is small, and that the plasma pressure is negligible (very small β). Thus, $B_z \cong B_{z0} = constant$. We also assume that $R/a \sim \epsilon^{-\frac{1}{2}}$ and $q \sim 1$. Using the approximation $\partial/\partial r \approx ik_r$ on equation (1), known as the

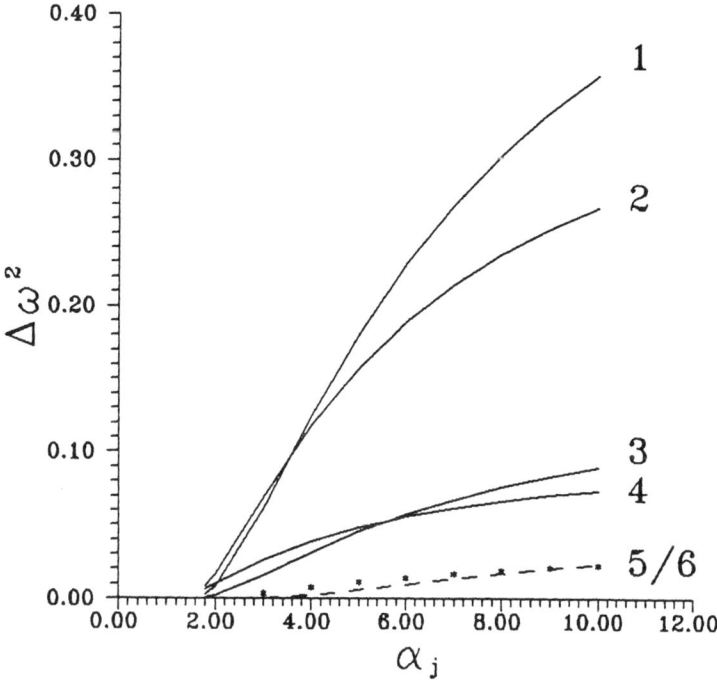

Fig. 1. Distance between the discrete Alfvén frequency and the lower edge of the Alfvén continuum, $\Delta\omega^2 = \omega_{DAW}^2 - \omega_{A0}^2$ versus α_j. The following input parameters are used for the curves:

no. curve	R	m, n	q_0	B_{z0}	
(1)	5	1, 2	1	2	
(2)	5	2, 2	1	1	
(3)	5	1, 2	1	1	
(4)	5	1, 1	1	1	
(5)	5	1, 2	2	1	(*)
(6)	10	1, 2	1	1	(- - - -)

WKB approximation, and expanding the dispersion relation up to the order one in ϵ, we derived the following dispersion relation:

$$\left(\omega^2 - \omega_{MSW}^2\right)\left(\omega^2 - \omega_{DAW}^2\right)\left(\omega^2 - \omega_{HLW}^2\right) = 0 \qquad (2)$$

where

$$\omega_{DAW}^2 = \omega_A^2 - \frac{[4k^2b^2 - 2kb'/rmB_z - Hr(b^2)']}{k_T^2\rho}, \qquad (3)$$

$$H = \frac{m^2}{r^2} + k^2 \cong \frac{m^2}{r^2}, \quad k_T^2 = H + k_r^2, b = \frac{B_\theta}{r} \tag{4}$$

and primes indicate the derivative with respect to r. The r-dependence is eliminated by replacing r by an effective radius r_{eff}. This effective radius is estimated to be of order of $0.1a$ to $0.2a$, deduced from the results of more complete calculation done numerically. The ω_{DAW}^2 represents the angular frequency of the discrete Alfvén wave[6].

NUMERICAL RESULTS

The Eq.(1) is solved by numerical technique using the shooting method. We can chose arbitrarily two profiles: pressure and B_θ. The pressure is given by $p(r) = p_0 \, exp(-4r^2 - 6r^4)$, where p_0 is constant. The B_θ profile is obtained by solving the Ampère's law with a given J_z profile, $J_z(r) = J_0 \left\{1 - r^2\right\}^{\alpha_j}$ where α_j is free parameters and the constant J_0 must satisfy the condition that the safety factor at the axis $q(0)$ is greater than 1. We choose a parabolic density profile, $\rho(r) = \rho_0 \left\{1 - 0.95r^2\right\}$. As a consequence of the inhomogeneous density we have a continuous spectrum of the Alfvén wave, in addition to discrete espectra. The shooting method is used to obtain the eigenvalue ω which satisfies the boundary condition at $r = 0$ and a. We choose the following parameters: $B_{z0} = 1$, $\rho_0 = 1$.

In figure (1) it is plotted the distance $\Delta\omega^2 = \omega_{DAW}^2 - \omega_{A0}^2$, where $\omega_{A0}^2 = k_\parallel^2 v_{A0}^2$, as a function of constant α_j, for different parameters. In this figure it is evident that the dependence analysed in the previous section. We can see that the discrete Alfvén wave only appear for α_j larger than 2. This means that the current profile should be very peaked to have a discrete Alfvén mode. For α_j smaller than 2 this mode enters the Alfvén continuum, that is, we do not have the discrete mode anymore.

CONCLUSIONS

We have shown that the discrete Alfvén wave exists in cylindrical plasma when the current profile is very peaked, $\alpha_j \geq 2$, where $J_z \sim (1 - r^2/a^2)^{\alpha_j}$. An approximate formula is derived to calculate the discrete Alfvén frequency for the case of small magnetic twist and very low β value. The most important result of this analysis is equation (3) which shows the dependence of the DAW eigen-angular frequency all with plasma parameters.

ACKNOWLEDGEMENTS

This work was supported by CNPq - Conselho Nacional de Desenvolvimento Científico e Tecnológico, and FAPESP - Fundação de Amparo a Pesquisa do Estado de São Paulo.

REFERENCES

1. P. Ulmschneider, E. R. Priest, R. Rosner, Proc. of the International Conference on Mechanisms of Chromospheric and Coronal Heating, Heidelberg, 5 - 8 june 1990 (Springer-Verlag, 1991).

2. J.P. Goedbloed, Phys. Fluids 18, 1258 (1975).
3. De Chambrier et al., Plasma Phys. 24, 893 (1982).
4. D.W. Ross et al., Phys. Fluids 25, 652 (1982).
5. K.Appert et al., Plasma Phys. 24, 1142 (1982).
6. C.A. de Azevedo et al., Solar Phys. 131, 119 (1991).
7. K. Hain and R. Lüst, Z. Naturforsch. 13a, 936 (1958).

THE X-RAY ULTRAVIOLET IMAGER FOR THE ORBITING SOLAR LABORATORY

Ester Antonucci[1], Marco Malvezzi[2], Luigi Ciminiera[3], Francesco Angrilli[4], Marilyn E. Bruner[5], Giovanni Perona[3], Maria Adele Dodero[1], Brian L. Evans[6], Leon Golub[7], Massimo Landini[8], Giancarlo Noci[8], Peter McWhirter[9], Brunella Monsignori Fossi[10], Giannina Poletto[10], Donald F. Neidig[11], Wolfgang K.H. Schmidt[12], Roger J. Thomas[13], Giuseppe Tondello[4].

ABSTRACT

A normal incidence multimirror telescope, the X-ray Ultraviolet Imager, for high resolution imaging of the solar atmosphere in the soft X-ray/XUV region, is being developed as part of the scientific payload of the NASA Orbiting Solar Laboratory. The X-ray Ultraviolet Imager is formed by two units: a high resolution telescope (0.25 arcsec pixel size and 8×8 arcmin^2 field of view) and a wide field one (2.3 arcsec pixel size and 5×5 solar radii^2 field of view). The two systems complement each other and allow a full coverage of solar features from the small scale (200 km on the sun) to the global phenomena. Each system consists of 8 channels with multilayer mirrors, imaging at different wavelengths. In each channel the mirror coating is optimized to select a narrow spectroscopic window corresponding to an intense line in the region 40 – 400Å. In order to provide imaging and temperature diagnostics from the chromosphere to the upper corona, 8 wavelengths are chosen to cover the broad temperature range from 10^5 to 10^7 K. Four images, two high resolution and two full disk ones, are simultaneously obtained by the X-ray Ultraviolet Imager, at a cadence which in flares can be of 0.4 – 1 s.

[1] University of Torino, Torino, Italy.
[2] University of Pavia, Pavia, Italy.
[3] Politecnico di Torino, Torino, Italy.
[4] University of Padova, Padova, Italy.
[5] Lockheed Palo Alto Research Laboratory, Palo Alto, CA, USA.
[6] University of Reading, Reading, UK.
[7] Smithsonian Astrophysical Observatory, Cambridge, MA, USA.
[8] University of Firenze, Firenze, Italy.
[9] Rutherford Appleton Laboratory, Chilton, UK.
[10] Osservatorio Astrofisico di Arcetri, Firenze, Italy.
[11] Phillips Laboratory (AFSC), National Solar Obs., Sunspot, NM, USA.
[12] Max Planck Insitute für Aeronomie, Lindau, Germany.
[13] NASA Goddard Space Flight Center, Greenbelt, MD, USA.

INTRODUCTION

The scientific payload of the Orbiting Solar Laboratory (OSL) includes the X-ray Ultraviolet Imager (XUVI), a normal incidence multimirror telescope. The concept of XUVI has been developed by a scientific team led by a group of italian scientists, supported by the Italian Space Agency in collaboration with the Phillips Laboratory of the US Air Force and the Max Planck Institute (Lindau, Germany). The X-ray Ultraviolet Imager is intended to complement and complete the long term monitoring of solar phenomena, at ultra high spatial and temporal resolution, performed by the other instruments mounted on OSL. This will be achieved by extending the observations from the infrared - visible - UV domain, covered by the Coordinated Instrument Package and the UV spectrograph, to the XUV and soft X-ray region.

XUVI is a normal incidence telescope with mirrors coated with multilayers; it is designed to provide excellent imaging properties, with sub-arcsecond spatial resolution (0.25 arcsec pixel size) in the 40 – 400Å spectroscopic region, coupled with high temporal resolution (0.4 seconds), although with moderate spectral resolution.

Lines emitted in the spectral range 40 – 400Å provide an extremely powerful diagnostic tool for solar system and astrophysical plasmas. In this region in fact there are many lines formed at greatly different temperature regimes (10^5–10^7 K).

Only recently, with the development of the new technology of reflecting surfaces coated with multilayers, it has become possible to detect the soft X-ray/XUV radiation emitted at a few 10^2Å with normal incidence telescopes, which have greatly improved the imaging capabilities in that spectroscopic region. In fact, the normal incidence reflectivity of metal coatings is very small ($R = 10^{-4} - 3 \times 10^{-2}$) at XUV wavelengths. However, considerably higher reflectivities, up to 0.3 – 0.4, can be obtained by using a mirror coated by depositing thin layers of suitable materials; since in this case, the reflectivity is enhanced by the in-phase reflections from the series of interfaces at the various layers.

So far excellent performances of multilayer mirrors, in the wavelength range 40 – 310Å, have been obtained in rocket flights [1,2,3]. The only telescope, mounted on a spacecraft, which used this technique, has been successfully flown on Phobos [4]. However, the loss of the spacecraft allowed only a limited test of its operational capabilities. The next imaging telescope based on multilayer optics planned for a long term mission is the Extreme-Ultraviolet Imaging Telescope, to be flown on SOHO in 1995.

Our present knowledge of the outer solar atmosphere mainly results from the success of previous space missions, such as the OSO series, Skylab and SMM. However, none of such missions has provided high resolution images by using radiation from different ions formed in a sufficiently broad temperature range to cover simultaneously the outer solar atmosphere, from the upper chromosphere, at a few 10^5 K, to the outer corona, at several million degrees. The only exception has been the high resolution imaging system of the NRL Skylab Spectroheliograph which however obtained partially overlapped images, by using a slitless system [5].

The X-ray Ultraviolet Imager is designed to overcome this serious deficiency in solar observations by imaging the different layers of the solar atmosphere, from the cool chromosphere up to the hot corona, thus providing not only the necessary link between the low and high atmospheric levels, but also the capability of studying simultaneously the wide variety of temperature regimes existing in the solar atmosphere. It will obtain narrow band images of the sun, in narrow spectral windows centered on prominent lines forming at different temperatures, to cover the temperature range from 10^5 K to 10^7 K. In this way, XUVI will produce unique measurements of temperature, emission measure and, with some limitations, density of the coronal plasma.

SCIENTIFIC OBJECTIVES

The present concept of XUVI evolves from a series of scientific objectives which include, in addition to the study of coronal heating which still remains one of the main open problems in solar physics, the investigation of the different processes and phenomena: from the small scale activity, to coronal loops and their activity, and to the large scale phenomena, such as coronal holes and coronal mass ejections (CMEs).

An effective progress in solar studies can only be achieved by improving the spatial and temporal resolution of observations up to the limit imposed by the present technology. A second requirement is to ensure a complete coverage of all structures in the solar atmosphere at the different heights and temperatures. In addition, in order to determine the context in which small scale activity, coronal loops and flares develop and to understand the impact on the global scale of smaller scale features, the capability of observing the full disk is also required.

The space era brought quite a change in our traditional picture of the sun. It was recognized that nearly all of the coronal emission originates in loop structures; since areas dominated by only one magnetic polarity, where open field lines are rooted, the so called coronal holes, are practically devoid of emission.

As space observations improved in spatial and temporal resolution, from OSO-7 (res. $\leq$ 20 arcsec), to Skylab (res. $\leq$ 5 arcsec) and HRTS (res. $\simeq$ 1 arcsec along the slit), to the latest rocket experiments like NIXT (res. $\simeq$ 3/4 arcsec), a wealth of small scale dynamic events has been progressively revealed.

The primary goal of XUVI is that of resolving small scale activity in order to give a substantial contribution to the problems of the heating of the solar corona, the origin and acceleration of the solar wind, and to the magnetic dynamo process.

The heating mechanism for the solar corona is elusive. The energy flux carried by acoustic waves is by far too small to provide for coronal heating [6,7]. Wave heating via the dissipation of Alfvén waves, which represents an obvious evolution of the acoustic wave theory, meets with major theoretical difficulties. On the other hand, Alfvén waves have not yet been detected in the solar corona. The accumulating evidence for a ubiquitous presence of small scale transient events has recently suggested a different approach to the heating problem. The impulsive energy release from small scale activity, may, via the superposition of a large number of events, produce the apparently steady phenomenon known as the solar corona [8].

Also the problem of solar wind acceleration may benefit from observations of small scale activity. OSO-7 and Skylab observations have shown that coronal holes are the sites where high speed solar wind streams originate. It is not clear, however, how the solar wind gets accelerated and if, and how much, small scale features contribute to the acceleration process. In this context UV jets have been proposed as a viable input of kinetic energy [9].

High resolution observations are also relevant for the understanding of the dynamo process. In the hypothesis that X-ray events are indicative of flux emergence, Golub [10] has shown that small events contribute to the total flux at least as much as large active regions. Because of the apparent importance for magnetic dynamo theories of flux emergence on small spatial scales, it is therefore essential to determine the limiting scale of the emergence process in order to define the magnetic spectrum.

Not only the understanding of the physics of small scale activity is expected to improve with solar imaging at high spatial resolution, but also that of coronal loops and related activity. In fact, observational data are still lacking both to identify the elementary loops and to develop definitive loop models.

The observed discrepancy between electron densities, derived by combining emission measure and geometrical information, and those determined by density sensitive line ratios, has been interpreted in terms of filling factors of the order of $10^{-4} - 10^{-2}$. These values would imply highly filamented flaring systems of loops with diameter of only $\simeq 150$ km [11], which can be resolved only with sub-arcsecond space resolution.

At least two different conditions in which loops can occur have been identified: a high temperature one ($T \geq 10^6$ K), where static conditions prevail, and a low temperature one ($T < 10^6$ K), characterized by plasma flows [12,13]. However, it is not clear on the basis of observations whether two classes of physically distinct features exist or whether cool loops simply represent the cool cores of hot loops, as suggested by Foukal [14,15].

In order to define the nature of coronal loops, it is therefore essential to identify the elementary loops, to determine their radial structure, and to verify the presence of discrete low level heating events showing up as a loop brightness variability.

One of the main problems to be addressed in flare studies is the question of energy storage. While it is generally agreed that magnetic fields are the source of flare energy, it is not yet clear under what circumstances loops form the high energy configurations leading to flares, nor has it definitely been shown that the magnetic field changes between pre- and post-flare structures can quantitatively account for the energy released [16]. The two most popular mechanisms which have been proposed to meet the flare energy requirements suggest either the storage of energy via the production of sheared loops as a result of some relative rotation of footpoints, or the emergence of magnetic flux tubes whose free energy is estimated to be higher than the energy accumulated by sheared loops and thus likely to be the source of the energy released by the largest flares [17]. Thus it is vital to observe the evolution of low beta plasma structures at high spatial resolution, in order to determine precisely the sequence of configurations leading to energy accumulation.

Progress in studies of energy release and transport in the flare impulsive phase relies on the capability of recording simultaneously several XUV lines with high temporal and spatial resolution. In fact, critical tests for models of energy release in flares, invoking either thermal heating or particle acceleration, rely on precise measurements of the relative timing between hard X-rays and other emissions originating in different temperature and density regimes of the atmosphere, as well as on the identification of the initial sites of enhanced emission and their relative positions.

A thorough study of large scale phenomena has not been possible in the past as previous space missions, OSO, Skylab and SMM, have not provided multiwavelength observations of the full disk, spanning all heights in the solar atmosphere, with high spatial resolution and over an extended period of time.

A global study of the corona makes it possible to estimate the total radiated power loss from the outer corona and its variation from feature to feature, e.g. active regions, coronal holes, quiet sun, etc.. This will be a valuable contribution to the understanding of the power balance in the atmosphere as a whole and the first time that such a comprehensive measurement will be possible.

Long term full disk monitoring of the corona should also provide data essential to understanding the underlying physics of CMEs and coronal holes. It is already known that CMEs are related to both flares and erupting prominences [18,19], but some CMEs appear to have no signature in the lower atmosphere of the sun, and it is not known, in general, whether the instability leading to the CME develops first in the corona or lower down. Until now, CMEs have been best studied by coronagraphs which can detect them only late in their development; therefore, very little is known about the physical circumstances leading to their formation.

Table I XUVI Performance requirements.

	HRI	FDI
Wavelength range	40 – 400 Å	40 – 400 Å
Temperature range	$8\ 10^4 - 2\ 10^7$ K	$8\ 10^4 - 2\ 10^7$ K
Spatial res.(pix. size)	0.25 arcsec	2.3 arcsec
Temporal resolution	≥ 1 s	≥ 0.4 s
Spectral resolution	14 – 100	10 – 50
Field of view	512×512 arcsec^2	4680×4680 arcsec^2

Skylab studies of the physical parameters of coronal holes have been hindered by the low sensitivity of the experiments. Better quality data are needed to derive the variation of parameters, such as temperature and emission measure, during the lifetime of coronal holes. So far, it has not been possible to correlate the coronal hole interior evolution with solar wind measurements.

OBSERVATIONAL REQUIREMENTS

In order to address the scientific objectives discussed above, XUVI has to meet several constraints. To image the finest solar features, from the transition region to the outer corona, an unprecedented spatial resolution coupled with high temporal resolution is required. As a consequence, XUV radiation originating over the entire temperature range of the solar atmosphere above the chromosphere, $10^5 - 10^7$ K, has to be simultaneously recorded. On the other hand, for the study of large scale phenomena, it is necessary to have a continuous monitoring of the full sun.

High spatial resolution at the sub-arcsec level as well as full disk monitoring cannot be fulfilled by the same imaging system. XUVI is therefore designed to include two optical systems, a high resolution imager (HRI) with a resolution of 0.25 arcsec (pixel size) on a limited field of view (8 × 8 arcmin^2), and a wide field telescope (Full Disk Imager, FDI) imaging the corona up to 2.5 R, with a spatial resolution of 2.3 arcsec (pixel size) adequate to study simultaneously large scale coronal structures and still most of the small scale solar phenomena. The field of view of the HRI is centered on that of the Coordinated Instrument Package of OSL, to ensure the capability of obtaining coordinated observations.

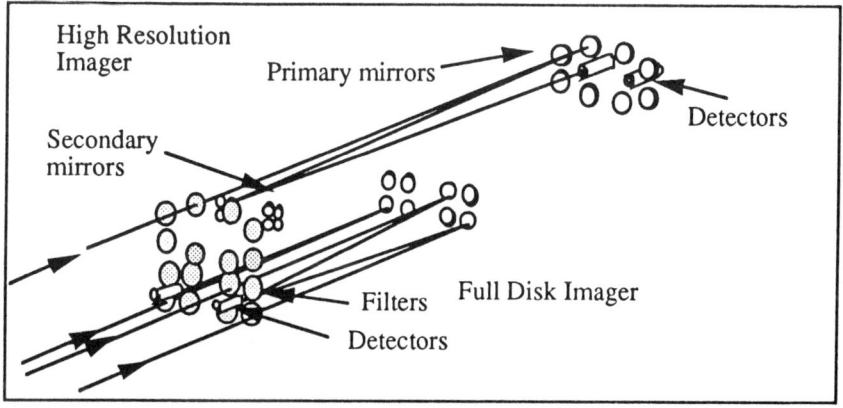

Fig. 1. Conceptual layout of the optical components of XUVI. The sun radiation enters from the lower left of the drawing. Four shutter wheels (not shown) located before the filters select the four active telescopes.

The performance requirements on both HRI and FDI telescopes are summarized in Table I.

The images formed by XUVI are not perfectly monochromatic due to the moderate spectral resolution, $\lambda/\Delta\lambda \simeq 10-100$. It is however possible to exploit the characteristics of the emission in the XUV spectral region in order to achieve in any case good diagnostic capabilities for plasma temperature.

The HRI and FDI are multi-mirror systems each consisting of eight channels providing narrow band images in different spectral windows, corresponding to intense XUV lines, in the range 40 – 400 Å. The number of spectral regions which can be observed in the XUVI channels is sufficient to cover the entire temperature range from 10^5 K to 10^7 K.

Table II   XUVI Optical characteristics.

	HRI	FDI
Primary mirror	spherical	spherical
Mirror diameter	10 cm	5 cm
Radius of curvature	14 m	4 m
Secondary mirror	plane	-
Focal length	7 m	2 m
Optical path	folded	-
Collecting area	78.5 cm^2	19.6 cm^2
Pixel size (on focal plane)	8.48 μ	22.17 μ

In each imaging system, the eight channels are clustered around two detectors which can sequentially record images from each channel. XUVI will therefore provide four simultaneous XUV images: two at high resolution and two of the full sun at different wavelengths. The instrument is designed to be equipped with 2048 × 2048 intensified CCD detectors.

The optical characteristics of the XUVI design are summarized in Table II and the technical specifications of the instrument are summarized in Table III. The overall optical design is illustrated in Figure 1.

Table III   XUVI Technical specifications.

Length	3.8 m
Diameter	0.6 m
Weight	170 kg
Power consumption	90 W
Telemetry	2 Mbps

The ultimate angular resolutions of the HRI and FDI are limited by the state of the art in optical manufacturing rather than by the geometrical aberrations of the optical system.

FDI can play a major role in the study of solar activity. In fact, previous space missions have demonstrated the difficulty of precise and timely repointing of telescopes with limited field of view on the sites of a flare at the time of its onset. This difficulty increases if the objective of the observation is the investigation of the origin of a CME. Therefore it is essential for such studies to have a wide field telescope, which continuously monitors the full sun, capable of detecting the pre-flare and pre-CME physical conditions in all the sites where these phenomena will develop.

The spectral region within 40 – 400Å allows temperature diagnostics at least between log T = 4.9 and log T = 7.2, sufficient to gain information on the chromosphere and low transition region, at the one extreme, and on flare

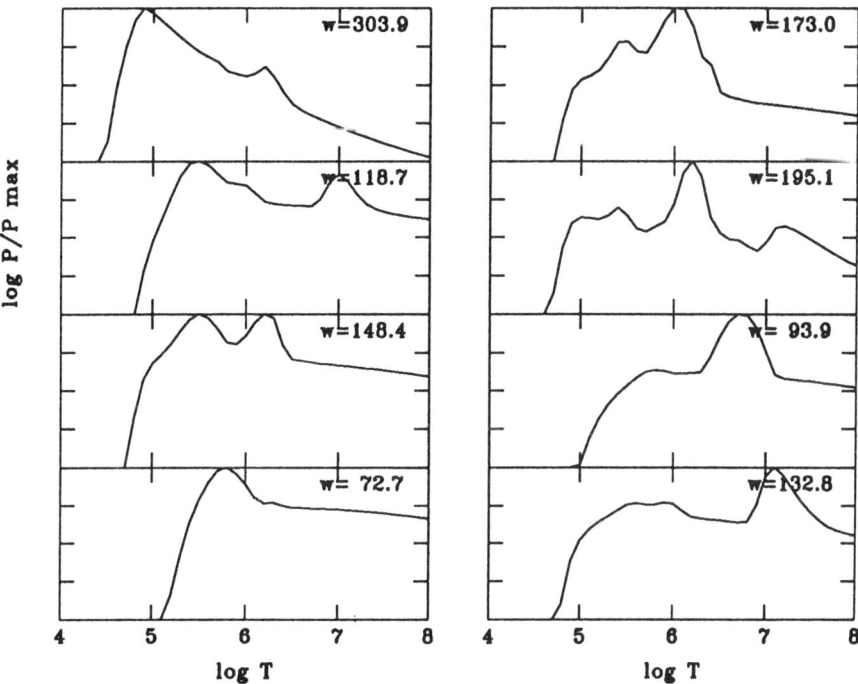

Fig. 2. Contribution functions of the XUV emission within the spectral band accepted by a multilayer mirror, with reflectivity optimized for the wavelength indicated in the figure. The functions are displayed according to increasing peak temperature. The units of quantity P are photons cm^{-3} s^{-1}; λ is the central wavelength of the spectral band.

conditions at the other. This spectral region includes many strong lines, which can assure counting rates sufficiently large for low density features or transient phenomena on high spatial resolution images.

In particular, this spectral region is rich in many very different stages of ionization of iron lines from Fe VIII to Fe XXII covering a temperature interval from log T = 5.9 to log T = 7.0, of oxygen and neon lines (from O IV to O VI and from Ne V to Ne VIII) covering temperatures from log T = 5.4 to log T = 5.8, lines of Si VIII, S VIII and S XI sensitive to the temperature interval log T = 6.4 - 6.9, and a few very high temperature lines (Fe XXIV) which can provide information on flare plasma (log T = 7.2).

The temperature coverage which can be achieved in the XUV region by a multilayer mirror telescope is illustrated by the contribution functions computed for different wavelengths, corresponding to some of the most intense lines in the XUV region (Figure 2). The computations have been performed by using the catalogue compiled by Landini and Monsignori-Fossi [20]. These contribution

134 The X-ray Ultraviolet Imager

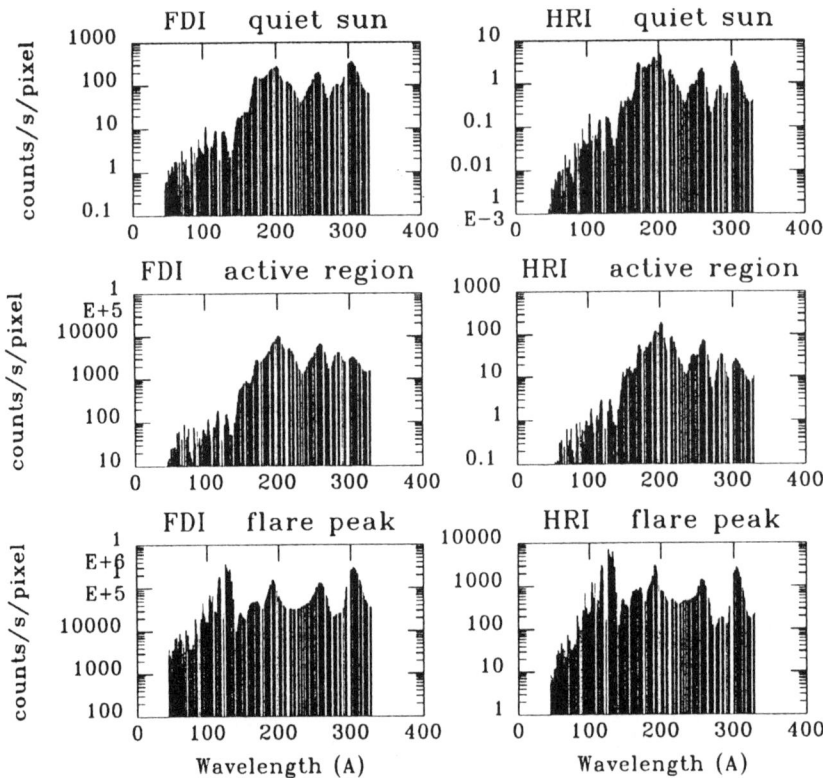

Fig. 3. Counting rates computed for the FDI and HRI, in the XUV spectral region, for the following solar conditions: quiet sun, active region, flare at the peak phase.

functions measure the emission in spectral windows whose width is determined by the properties of the multilayer which optimizes the reflectivity at a given wavelength.

Expected counting rates in the FDI and HRI have been computed for the most prominent XUV spectral lines, in the spectral window determined by the mirror and centered on the line.

The FDI and HRI counting rates (as a function of wavelength in Figure 3) have been derived for the following solar conditions: quiet sun, active region and flare peak. In the computation, the differential emission measure derived by Gabriel [21], Withbroe [22] and Dere and Cook [23] have been used, respectively.

With a suitable choice of wavelengths for the imager channels, high temporal resolution (~ 1 s) can be achieved with the HRI in active regions and flares. Full disk images of the quiet sun can be obtained continuously with the FDI at a temporal resolution of 10 seconds, while in detecting active regions and during flares the integration time can be as short as 0.2 seconds.

REFERENCES

1. J.H. Underwood, M.E. Bruner, B.M. Haisch, W.A. Brown, and L.W. Acton, Science 238, 61 (1987).
2. A.B.C.Jr. Walker, T.W.Jr. Barbee, R.B. Hoover, and J.F. Lindblom, Science 241, 1781 (1988).
3. L. Golub, M. Herant, K. Kelata, I. Lovas, G. Nystrom, F. Pardo, E. Spiller, J. Wilczynsky, Nature 344, 842 (1990).
4. I.I. Sobel'man, et al., D.N. Lebedev Physical Institute Preprint, Moscow 241, 1 (1988).
5. K.P. Dere, Solar Phys. 77, 77 (1982).
6. R.C. Athay, O.R. White, Astrophys. J. 229, 1147 (1979).
7. E.C.Jr. Bruner, Astrophys. J. 247, 317 (1981).
8. E.N. Parker, Astrophys. J. 330, 474 (1988).
9. G.E. Brueckner, J.-D.F. Bartoe, Astrophys. J. 272, 329 (1983).
10. L. Golub, Phil. Trans. R. Soc. Lond. 297, 595 (1988).
11. G.A. Linford, C.J. Wolfson, Astrophys. J. 331, 1036 (1988).
12. C.-C. Cheng, J.B. Jr. Smith, E.A. Tandberg-Hanssen, Solar Phys. 67, 870 (1980).
13. S.K. Antiochos, G. Noci, Astrophys. J. 301, 440 (1986).
14. P. Foukal, Astrophys. J. 210, 575 (1976).
15. P. Foukal, Astrophys. J. 223, 1046 (1978).
16. R.L. Moore, Astrophys. J. 324, 1132 (1988).
17. A.N. McClymont, G.H. Fisher, Solar System Plasma Physics, Geophysical Monograph 1, 219 (1989).
18. R.A. Harrison, Astron. Astrophys. 162, 283 (1986).
19. D.F. Webb, A.J. Hundhausen, Solar Phys. 108, 383 (1987).
20. M. Landini, B.C. Monsignori Fossi, Astron. Astroph. Suppl. Ser. 82, 229 (1990).
21. A.H. Gabriel, Phys. Trans. R. Soc. Lond. A. 281, 339 (1976).
22. G.L. Withbroe, Solar Phys. 45, 301 (1975).
23. K.P. Dere, J.W. Cook, Astrophys. J. 229, 772 (1979).

HYDROSTATIC MODELS OF X-RAY CORONAL LOOPS OBSERVED BY NIXT

G. Peres
Osservatorio Astrofisico di Catania, Catania, Italy

F. Reale
IAIF/CNR, Osservatorio Astronomico, Palermo, Italy

L. Golub
Center for Astrophysics, Cambridge (MA), USA

ABSTRACT

Observations made with the Normal Incidence X-ray Telescope (NIXT) have shown that some X-ray emitting structures observed with NIXT resemble very closely the corresponding Hα plages. We have used hydrostatic models of coronal loops to explain such observations as strong emission from the lower section of high-pressure coronal arches.

INSTRUMENT AND OBSERVATIONS

The Normal Incidence X-ray Telescope[1] (NIXT) is an f/8 Ritchey Chretien with diameter 25 cm and is based on multilayer mirrors; the spectral response is centered on 63.5Å so as to detect the intense Mg X and Fe XVI soft X-ray lines and the passband is 1.4Å; the angular resolution is below 1 arcsec and has been limited so far by the resolution of the photographic film used as detector.

NIXT has already been flown on a rocket, collecting on photographic film data of unprecedented high resolution in soft X-rays and characterized by very low scatter[2]. Coronal arches first shown by previous observations made with grazing incidence X-ray telescopes[3] are now shown with a much higher detail by the images taken with NIXT. Images show very fine details, in particular the structuring and filaments of the corona, and show virtually the whole range of coronal structures, including active regions and flaring loops. A novel feature of these images is the close resemblance of some structures detected by NIXT and the corresponding Hα plages, in addition to the usual loop-like structures typically shown by wide band X-ray images of the corona. We can safely exclude that such new features might be due to leakage of strong EUV emission, for instance, of the strong He 304Å line, because of the presence of a carbon filter in the optical path and because the film employed is not sensitive in such a band. Therefore we have tried to explain NIXT observations by means of coronal emission from "standard" hydrostatic models of coronal loops.

HYDROSTATIC MODELS AND RESULTS

The data from NIXT do not show significant variability of the structures over the observation duration of ∼ 10 minutes. Since other observations of the corona have shown that many loops remain unaltered for periods larger than the radiative and conductive cooling times, we assume that these structures can be described by hydrostatic models. We consider one-dimensional hydrostatic models based on the equations of energy and pressure balance[4], including the effects of conduction and radiative losses in corona and assuming a semicircular loop geometry, symmetric with respect to the apex:

$$-\frac{d F_c}{ds} - \beta n^2 P(T) + H = 0 \qquad (1)$$

$$F_c = -\kappa T^{5/2} \frac{dT}{ds} \qquad (2)$$

$$\frac{dp}{ds} = -\rho\, g(s) \qquad (3)$$

$$\rho = \mu\, m_H\, n \qquad (4)$$

$$p = (1 + \beta)\, n\, K_B\, T \qquad (5)$$

where s is the coordinate along the loop, F_c the conductive flux, β the ionization fraction, n the plasma particle density, $P(T)$ the radiative losses function from an optically thin plasma, H a steady heating assumed uniform all over the loop, g the component of the solar gravity along the field lines, T the plasma temperature, κ the conduction coefficient, according to the Spitzer[5] formulation, p the thermal pressure, μ the plasma specific mass per unit particle, m_H the proton mass, K_B the Boltzmann constant. The lower boundary of the loop is set at 20000K.

Table I Maximum Temperature ($\times 10^6$K) of Static Loop Models

L_8^a	5	15	50	150	500
p^b					
0.03	0.32	0.51	0.75	1.05	1.51
0.1	0.5	0.76	1.12	1.57	2.24
0.3	0.71	1.07	1.60	2.26	3.28
1	1.04	1.58	2.39	3.41	5.04
3	1.48	2.28	3.48	5.05	7.41
10	2.19	3.42	5.25	7.59	11.2
30	3.18	4.99	7.54	11.0	16.8

a - loop semilength (in units of 10^8cm)
b - loop base pressure (dyne cm^{-2})

In order to ascertain how a given loop would appear in a NIXT observation, we have computed the distribution emission in the NIXT band along the loop from the distributions of temperature and density obtained from the above model. The folding of the coronal plasma emissivity through the NIXT response yields a peak at $\sim 1 \times 10^6$K and another, lower by half a decade, at $\sim 3 \times 10^6$K, with a tail extending to higher temperatures[6]. We have performed this synthesis for a whole grid of loop models encompassing most of the typical conditions encountered in corona. In table I we report the models examined, characterized by the loop semilength and base pressure, and a corresponding temperature at the loop apex.

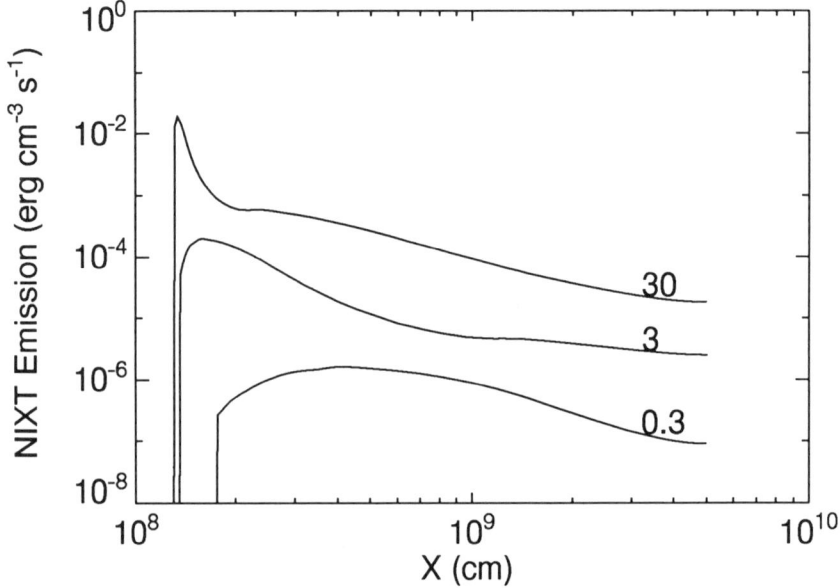

Fig. 1. Distribution of the emission in the NIXT band along the field line coordinate for hydrostatic models of loops with semilength $L = 5 \times 10^9$ cm. Base pressures are labelled (dyne cm^{-2}).

In figure 1 we report a representative subset of our results: the distribution of emission in NIXT band along the field line coordinate in loops of semilength $L = 5 \times 10^9$ cm and for various values of the base pressure, from 0.3 to 30 dyne/cm^2. The emission along the loop considerably increases with base pressure and moreover, as base pressure increases, the emission peak moves down towards the base of the loop. Above 1 dyne/cm^2 the emission peak is immediately above the base. Such dependence on pressure occurs for all the loop lengths examined.

In loops with still higher pressure the emission peak becomes a very prominent spike. We believe that this enhancement of emission at the base might explain the structures similar to Hα plages detected by NIXT. In other words, in such regions NIXT detects very bright loop footpoints, which therefore have close correspondence to the Hα plages located immediately below. The loop base is so bright because there the plasma is much denser than in the higher corona, and at temperatures in the range of maximum NIXT response.

Indeed this might be a new potential diagnostic tool, which could allow us to identify easily high pressure regions (where the emission enhancement preferentially occurs) from NIXT pictures. In this perspective the reality and the extent of this similarity between X-rays and Hα in these regions should be clarified in depth and, more specifically, detailed morphological studies of NIXT observations and extensive comparisons with specific models are needed.

REFERENCES

1. E. Spiller, R. McCorkle, Wilczynski J., L. Golub, G. Nystrom, P. Takacz and C. Welch, Proc. SPIE 1343, 134 (1990).
2. L. Golub, M. Herant, K. Kalata, I. Lovas, G. Nystrom, F. Pardo, E. Spiller and J. Wilczynski, Nature 344, 842 (1990).
3. G. S. Vaiana, A. S. Krieger and A. F. Timothy, Sol. Phys. 32, 81 (1973).
4. S. Serio, G. Peres, G. S. Vaiana, L. Golub and R. Rosner, Ap. J. 243, 288 (1981).
5. L. Spitzer, The Physics of Fully Ionized Gases (Interscience, N.Y., 1962).
6. L. Golub and M. Herant, Proc. SPIE 1160, 629 (1989).

THE EFFECT OF VISCOSITY ON HYDRODYNAMICS OF CORONAL FLARES

F. Reale
IAIF/CNR, Osservatorio Astronomico, Palermo, Italy

G. Peres
Osservatorio Astrofisico di Catania, Catania, Italy

ABSTRACT

We investigate the effect of plasma viscosity in the hydrodynamics of coronal flares. To this end we compute two otherwise identical models of a typical coronal compact flare, one including and the other neglecting viscosity terms from the relevant hydrodynamic equations. We find significant differences which may affect the diagnosis of observed high resolution X-ray spectra.

INTRODUCTION

Hydrodynamic models of coronal plasma evolution have been extensively used to interpret solar and stellar observations of X-ray flares[1,2], including high resolution X-ray spectra[3].

Given the increasing importance of high resolution spectral analysis, we are investigating the influence of plasma classical viscosity[4] in flare models both on the plasma dynamics, and particularly on the ensuing spectra. We have considered a one-dimensional hydrodynamic model of compact flares within a semicircular loop and we have made two otherwise identical calculations of the impulsive phase of a solar flare, one including and the other omitting viscosity from the relevant hydrodynamic equations.

We have found significant differences in the velocity distribution and evolution and also in the synthesized high resolution spectra of Ca XIX line (3.18Å), as synthesized from the two models.

THE MODELS

In the course of this study we have used the Palermo-Harvard 1-D hydrodynamic code, which has been extensively applied to previous solar and stellar studies[1,2,5,6,7].

The model describes the evolution of plasma confined in semicircular coronal loops of constant cross-section. The symmetry assumed allows us to describe only half coronal loop. The initial atmosphere is hydrostatic and includes a corona and a chromosphere, connected by a steep transition region[8].

The model is based on the standard non-linear differential hydrodynamic equations of mass, momentum and energy conservation in one dimension:

$$\frac{dn}{dt} = -n\frac{\partial v}{\partial s}$$

$$nm_H \frac{dv}{dt} = -\frac{\partial p}{\partial s} + nm_H g + \frac{\partial}{\partial s}\left(\mu \frac{\partial v}{\partial s}\right)$$

$$\frac{d\mathcal{E}}{dt} + w\frac{\partial v}{\partial s} = \mathcal{Q} - n^2 \beta \mathcal{P}(T) + \mu \left(\frac{\partial v}{\partial s}\right)^2 + \frac{\partial}{\partial s}\left(\kappa \frac{\partial T}{\partial s}\right)$$

$$p = (1+\beta) n k_B T$$

$$\mathcal{E} = \frac{3}{2} p + n\beta \chi$$

$$w = \frac{5}{2} p + n\beta \chi$$

where n is the Hydrogen number density, s the field line coordinate, v the plasma velocity, T the temperature, p the pressure, g the component of gravity parallel to field line, m_H the mass of Hydrogen atom, μ the viscosity, $\beta = n_e/n$ the ionization fraction, n_e the electron density, κ the thermal conductivity, χ the Hydrogen ionization potential, k_B the Boltzmann constant, $\mathcal{P}(T)$ the radiative losses per unit emission measure[9], $\mathcal{Q}(s,t)$ the volumetric power input to the solar atmosphere assumed to be the sum of a steady heating term, which maintains the equilibrium of the initial atmosphere, and a transient heating, which triggers the flare and can be formulated in different ways.

Viscosity terms are present both in momentum and energy equations. The two sets of simulations performed differ only in the viscosity coefficient μ in corona: in one case the coefficient is the classical Spitzer[4] coefficient, in the other case it is set to zero. We have considered rather realistic flare parameters, which had been used in a previous detailed study of a compact flare observed by SMM[1]. The heating function is parameterized as a separable function of space and time: the spatial dependence is Gaussian with center at the top of the loop, a maximum value of 10 erg cm^{-3} s^{-1} and a width of 5×10^8 cm. As for time dependence, the heating is kept constant for 3 minutes and then decays exponentially with a characteristic time of 1 min.

RESULTS

We find significant differences during the first phase between the velocity profiles of the two models and also some difference in the density distribution at various times. The temperature is largely dominated by heat conduction and then the two models show no remarkable difference in the temperature profiles.

In figure 1 we show distributions of plasma velocity sampled at some representative times during the initial phase of the flare calculated with and without viscosity. The evolution shows the well known evolution leading to chromospheric evaporation: in the first 10s the impulsive heating rapidly enhances the temperature at the loop apex and a strong conduction front moves downwards, reaching the chromosphere in few seconds. The dense chromosphere is abruptly heated and rapidly expands upwards, with high positive velocity, visible in the low corona after 20s. The evaporation front reaches the top of the loop with

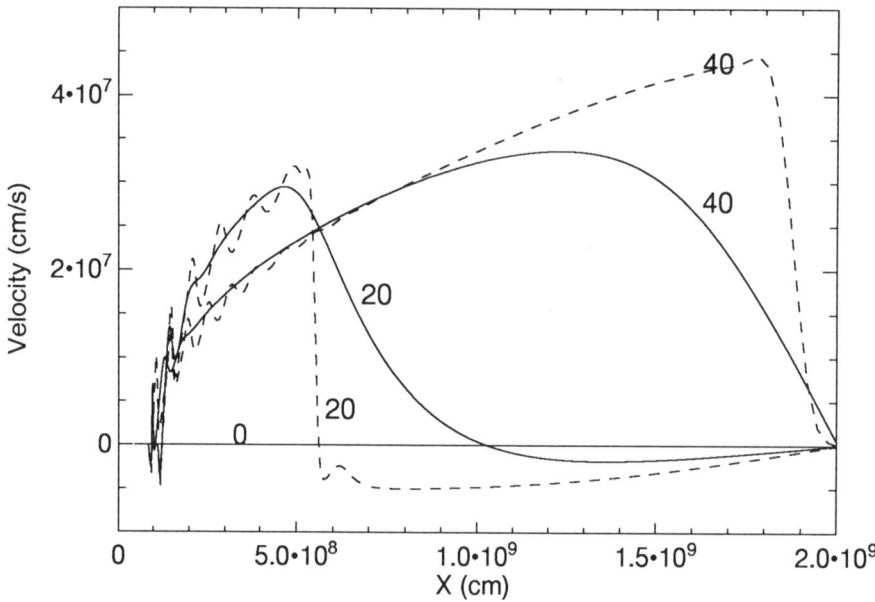

Fig. 1. Velocity distributions along the field line coordinate sampled at the labelled times (s) for the models with (solid) and without (dashed) viscosity.

velocities comparable to the speed of sound (~ 300km/s), and the loop is therefore progressively filled up with denser plasma. After 1 minute, as the loop approaches a new equilibrium state at a much higher pressure, upward motions become less conspicuous.

We want now to focus on the comparison between the calculations with and without viscosity. As shown in figure 1, a well-defined shock front is immediately apparent in the case without viscosity; such shock is strongly smoothed out in the case with viscosity where the velocity field is much more uniform. In the case without viscosity the small ripples are due to numerical noise and the peak velocity is quite higher (~ 25%) in the case without viscosity.

We have found that, in the absence of viscosity, there is static plasma in the upper regions of the loop in front of the shock front, even at later times, while this does not occur in the presence of viscosity. Incidentally, we note that these shock fronts have represented the major difference between the results of the present code and those of other codes which don't include viscosity[10,11,12].

We can expect that the significant differences detected in the velocity distributions may reflect on observational signatures, especially for high resolution spectra in soft X-ray lines, wherever plasma motions may produce significant

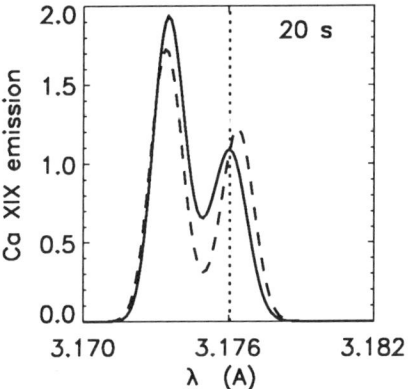 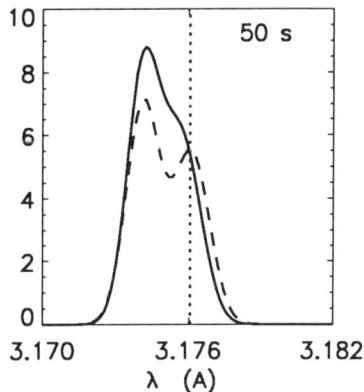

Fig. 2. Synthesized Ca XIX w line profiles (10^{14} erg cm^{-3} s^{-1} ster^{-1}) sampled after 20s and 50s since the beginning of the heat pulse, for the models with (solid) and without (dashed) viscosity. The vertical dotted line marks the unshifted line wavelength.

Doppler-shifted components.

We have calculated a high resolution spectrum of the w resonance line of Ca XIX helium-like ion line at 3.176Å We have used the formulation of Withbroe[13] and included the thermal broadening and the Doppler-shifts due to bulk plasma motions. We have considered a point of view from above the loop.

As an example we show in figure 2 two snapshots of Ca XIX resonance line synthesized from calculations with and without viscosity. A strong blue-shifted component is present in both panels, associated with upward chromospheric evaporation. Blue-shifts correspond to velocities of the order of 250km/s, much less than the peak velocity inferred from the velocity distributions shown in figure 1, because this result is obtained from integration on the whole loop.

Differences in the line profiles between cases with and without viscosity are quite significant at time t = 50s. The present results are preliminary ones and a more complete analysis is in progress; indeed they are already enough to warn about the important role that viscosity may have in flare hydrodynamic models, especially when used for diagnostics of high resolution spectra.

REFERENCES

1. G. Peres, F. Reale, S. Serio and R. Pallavicini, Ap. J. 312, 895 (1987).
2. F. Reale, G. Peres, S. Serio, R. Rosner and J. H. M. M. Schmitt, Ap. J. 328, 256 (1988).
3. E. Antonucci, M. A. Dodero, G. Peres, S. Serio and R. Rosner, Ap. J. 322, 522 (1987).

4. L. Spitzer, The Physics of Fully Ionized Gases (Interscience, N.Y., 1962).
5. G. Peres, R. Rosner, S. Serio and G. S. Vaiana, Ap. J. 252, 791 (1982).
6. S. Serio, F. Reale, J. Jakimiec, B. Sylwester and J. Sylwester, A & A 241, 197 (1991).
7. J. Jakimiec, B. Sylwester, J. Sylwester, S. Serio, G. Peres and F. Reale, A & A in press, (1991).
8. S. Serio, G. Peres, G. S. Vaiana, L. Golub and R. Rosner, Ap. J. 243, 288 (1981).
9. R. Rosner, W. H. Tucker and G. S. Vaiana, Ap. J. 220, 643 (1978).
10. C.-C. Cheng, E. S. Oran, G. A. Doschek, J. P. Boris and J. T. Mariska, Ap. J. 265, 1090 (1983).
11. J. T. Mariska, A. G. Emslie and Peng Li, Ap. J. 341, 1067 (1989).
12. G. H. Fisher, R. C. Canfield and A. N. McClymont, Ap. J. 289, 414 (1985).
13. G. Withbroe, Activity and Outer Atmosphere of the Sun (SAAS-FEE, 1981), p. 9.

ON NON-LOCAL TRANSPORT PROCESSES IN THE SOLAR ATMOSPHERE

P. MacNeice
NASA/GSFC, Greenbelt, MD 20771, USA

ABSTRACT

We review two mechanisms which can lend a non-local character to energy transport in the solar atmosphere, heat flux propagating in the form of collisionless electrons, and non-equilibrium ionization of hydrogen driven by ambipolar diffusion. Application of these processes to modelling of the lower transition region and upper chromosphere is considered.

INTRODUCTION

Transport processes with a non-local character have become increasingly popular of late as potential solutions to some longstanding problems in modelling the outer solar atmosphere. In this paper we will focus on two particular mechanisms, non-local heat flux arising from energetic collisionless electrons and non-equilibrium ionization driven by non-zero fluid velocities. We shall discuss these mechanisms in the context of a particular well defined problem, namely that of explaining the differential emission measure deficit observed in the solar transition region.

The central issue is to construct a model of the temperature and density structure in the solar plasma responsible for producing transition region and upper chromospheric emissions. The differential emission measure

$$\xi(T) = \int_{S_T} dS \, n_e^2(T) \left| \frac{T}{\boldsymbol{\nabla} T} \right| \qquad (1)$$

is a convenient measure with which to characterise this model because for any optically thin emission line, the contribution from the gas at temperature T is directly proportional to $\xi(T)$. S_T is all surfaces at temperature T throughout the source volume and n_e is the electron number density. When a sufficiently complete set of optically thin emission line intensities has been assembled an 'observed' differential emission measure can be obtained, assuming ionization equilibrium, by solving the associated inversion problem. The typical 'observed' differential emission measure is illustrated in figure 1. It has a minimum at roughly 10^5 K.

The simplest models have attempted to explain these observations in terms of quasi-stationary coronal loops. In these models the coronal plasma cools principally by magnetic field-aligned thermal conduction to the thin transition region and underlying chromosphere which radiate much more effectively than the corona. The thermally conducted heat flux is adopted from the classical

$$q = -\kappa_o T^{5/2} \boldsymbol{\nabla}_{\|} T \qquad (2)$$

formula first derived by Spitzer and Härm[1], where κ_o is a constant and $\boldsymbol{\nabla}_{\|}$ denotes the gradient in the direction of the magnetic field. The difficulty arises because the heat flux carried downward by conduction through the 10^5K temperature regime cannot all be radiated away by material at this temperature. However for conduction to transport this flux to lower temperatures requires the temperature gradient to steepen dramatically to counteract the falloff in the $T^{5/2}$ coefficient, and this steepening is inconsistent with the observed differential emission measure, as shown in figure 1.

Many suggestions have been made in an effort to improve this loop model. Gabriel[2] suggested allowing for convergence of the field lines defining the loop which would serve as a bottleneck and act to reduce the downward heat flux load. Pneumann and Kopp[3] and Athay[4] have suggested that the inclusion of downward mass flows and the associated enthalpy flux would reduce the amount of energy required to be transported downward by thermal conduction. Woods et al[5] have suggested that the wave pressure associated with a wave heating mechanism should be included in the momentum balance. However none of these suggestions has resolved the problem.

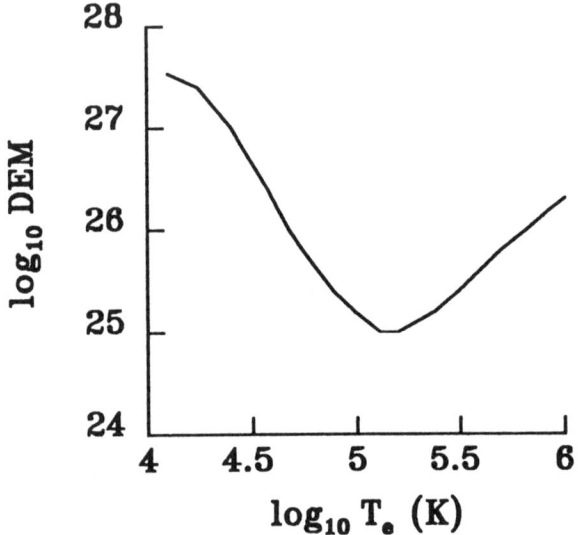

Fig. 1. A typical differential emission measure determined from observations (cf. Raymond and Doyle[6]).

A number of papers[7,8] have argued that separate loop structures which do not extend to coronal temperatures may be responsible for the dominant contribution to the lower transition region emissions. Lately, high resolution images in the EUV have provided some observational support for this viewpoint. However for our present purpose of discussing transport processes it serves us better to concentrate on new physical processes relevant to the single coronal loop model.

Shoub[9] suggested that in the presence of such steep temperature gradients, electrons originating in the hotter corona might create a collisionless energetic tail to the electron velocity distribution in the lower transition region. This could supplement the Spitzer-Härm heat flux thereby relaxing the classical relationship between temperature gradient and heat flux.

Finally Fontenla et al(FAL)[10] have proposed that non-equilibrium ionization of hydrogen, driven by ambipolar diffusion, may alter the concentrations of neutral hydrogen as a function of height, significantly altering the total energy flux by the addition of a flux of ionization energy, and also changing the location of the strong Ly α energy sink.

It is recent work concerning these last two suggestions which we review in this paper.

LOOP MODELS

Before discussing these two processes further we shall take a moment to introduce the loop models with which these hypotheses have been tested. These are a set of models of the upper chromosphere and lower transition region, computed by Fontenla et al[10], which have been extended to coronal temperatures by the addition of an overlying semi-circular loop in hydrostatic equilibrium.[11]

The FAL models differ from the semi-empirical models of Avrett[12] through their inclusion of ambipolar diffusion which they show becomes important in the upper chromosphere and lower transition region, and more accurate treatment of heat transport by thermal conduction in a partially ionized hydrogen plasma. Since neither thermal conduction nor ambipolar diffusion is important in the lower chromosphere and since the mechanical heating process there is poorly understood, the temperature structure below 8×10^3K in these models was assumed to be the same as that determined semi-empirically by Avrett[12]. Above 8×10^3K the temperature structure was determined by integrating the energy balance equation upward from the height at which $T = 8 \times 10^3$K occurs. This equation included energy transport by radiation and thermal conduction, transport of ionization energy and enthalpy by ambipolar diffusion and a simple mechanical heating rate. Its solution was obtained in conjunction with consistent solution of the equations of hydrostatic and statistical equilibrium and of radiative transfer. FAL followed the convention introduced by Vernazza et al[13] and repeated by Avrett[12] in constructing separate models considered more appropriate for the average quiet sun(C), a plage region(P), a very bright network element(F) and a dark point within a cell(A). The main differences between the transition regions in these models are that the increasing brightness is correlated with higher transition region pressure and steeper temperature gradient.

The extension of the model to coronal temperatures was achieved by integrating the equations for hydrostatic equilibrium and steady-state energy balance from the top of the FAL model to the loop apex. The heat flux q was chosen to be zero at the apex. The energy balance was achieved between optically thin radiative losses, field-aligned thermal conductivity and a simple heating function. This heating function was the sum of the FAL heating term, directly proportional to the mass density, and a weak component with gaussian spatial distribution centered about the loop apex as required to achieve $q(apex) = 0$. The thermal structure of model C is shown in figure 2.

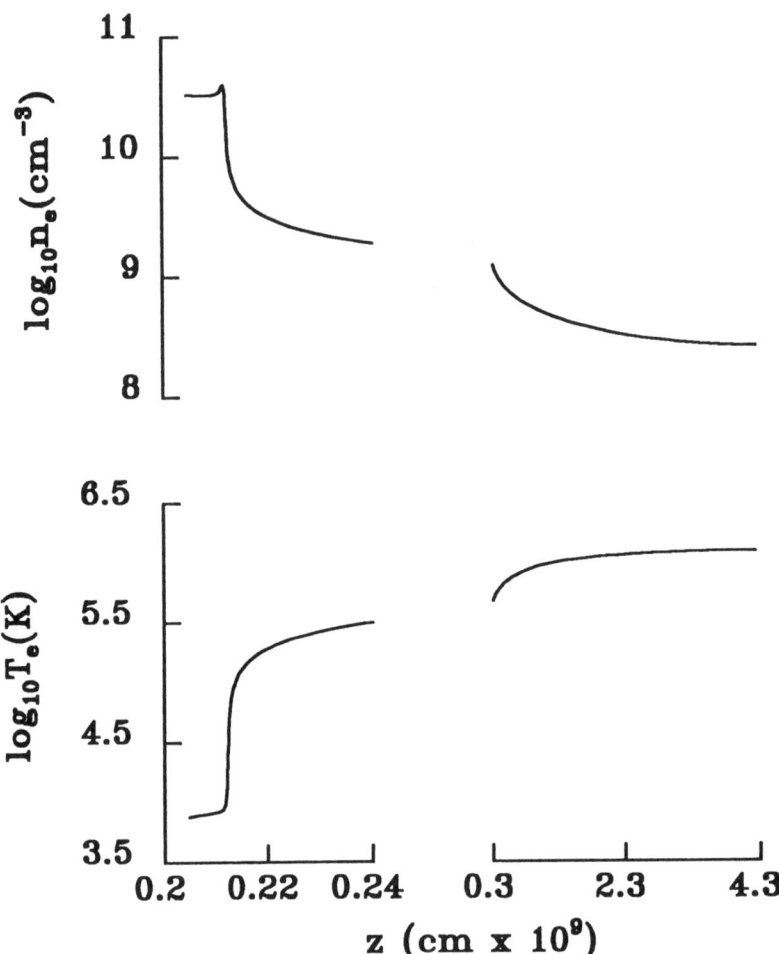

Fig. 2. The electron temperature and density profiles for model C and its coronal extension. The loop is assumed symmetric about its apex and so only one half is shown. The left frame illustrates the variation through the chromosphere and lower transition region while the right frame which has a coarser spatial resolution illustrates the upper transition region and coronal variation.

COLLISIONLESS HEAT FLUX

In an ionized plasma in which the principal scattering mechanism for electrons is the accumulated effects of 2-body coulomb collisions, the electron mean free path λ is proportional to $|v|^4$ where v is the electron velocity. Very high energy electrons therefore can travel many temperature scale heights $L = |T/\nabla T|$

between collisions. If collisionless electrons contribute significantly to the heat flux it acquires a non-local character. To compute the heat flux we must first determine the electron velocity distribution.

This is determined from the Landau equation[14] which can be written as

$$\frac{\partial f}{\partial t} + \boldsymbol{v} \cdot \frac{\partial f}{\partial \boldsymbol{x}} - e\frac{\boldsymbol{E}}{m_e} \cdot \frac{\partial f}{\partial \boldsymbol{v}} = \left(\frac{\partial f}{\partial t}\right)_c \quad (3)$$

where f is the electron distribution function, $\boldsymbol{x}$ is the position vector and $\boldsymbol{E}$ the electric field The right hand-side of the equation describes the rate of change of f as a result of collisional redistribution of electrons in velocity space, and is given by

$$\left(\frac{\partial f}{\partial t}\right)_c = -\Gamma \frac{\partial}{\partial v^\mu}\left(f\frac{\partial H}{\partial v^\mu}\right) + \frac{1}{2}\frac{\partial^2}{\partial v^\mu \partial v^\nu}\left(f\frac{\partial^2}{\partial v^\mu \partial v^\nu}G\right), \quad (4)$$

with

$$H(\boldsymbol{v}) = \sum_b \frac{m_e + m_b}{m_b} \int d^3v' \, f_b(\boldsymbol{v}') \mid \boldsymbol{v} - \boldsymbol{v}' \mid^{-1}, \quad (5)$$

$$G(\boldsymbol{v}) = \sum_b \int d^3v' \, f_b(\boldsymbol{v}') \mid \boldsymbol{v} - \boldsymbol{v}' \mid, \quad (6)$$

and

$$\Gamma = \frac{4\pi e^4 \ln\Lambda}{m_e^2}. \quad (7)$$

Here $\ln\Lambda$ is the Coulomb logarithm and b denotes the particle species in the plasma (including other electrons) with which the electrons collide. Obviously equation (3) is non-linear. Since we are only considering time-independent models here, we will drop the time derivative $\partial f/\partial t$.

Classical heat flux theory assumes a strongly collisional plasma in which only weak departures from maxwellian are permitted. Spitzer and Härm[1] calculated the time-independent electron distribution function which results from the presence of a weak electric field and temperature gradient. Their solution was based on a perturbation analysis of the Landau equation effectively expanding the distribution function as a power series in the Knudsen parameter $K = \lambda/L$ about a local Maxwellian. Since K is an increasing function of electron velocity, the Spitzer-Härm distribution (hereafter denoted by SH) always breaks down at sufficiently high electron velocity. For weak electric fields and temperature gradients this breakdown does not significantly compromise the ability of the SH solution to describe the transport properties of the plasma. However it has been shown[15] that for fully ionized plasma, this is no longer the case when K reaches values of ≥ 0.02. Then the heat flux may receive a substantial contribution from electrons which are collisionless, and whose behaviour is poorly described by the SH theory.

A more accurate description of the high velocity particles than the SH solution is provided by the solution of the high velocity form of the Landau equation (HVL),

$$v\mu\frac{\partial f}{\partial z} - \frac{e\boldsymbol{E}}{m_e} \cdot \frac{\partial f}{\partial \boldsymbol{v}} = \frac{1}{v^2}\frac{\partial}{\partial v}\left[v^2\nu(v)\left(\frac{kT_e}{m_e}\frac{\partial f}{\partial v} + vf\right)\right] - \nu(v)\frac{\partial}{\partial \mu}\left[(1-\mu^2)\frac{\partial f}{\partial \mu}\right], \quad (8)$$

where
$$\nu(v) = \frac{4\pi e^4 n_e \ln\Lambda}{m_e^2 v^3}. \tag{9}$$

This is a linearisation of the Landau equation based on the assumption that high velocity electrons are sufficiently uncommon that their scattering is dominated by collisions with slow electrons rather than with other high velocity electrons. In deriving equation (8) we have assumed that the plasma is homogeneous except in the direction of the s axis, which we will identify with the direction of the magnetic field, and $\mu = \boldsymbol{v} \cdot \hat{\boldsymbol{z}}/v$ is the cosine of the electron pitch angle.

By solving this equation Shoub[9] calculated the distribution of high velocity electrons for an atmospheric model with unrealistically steep temperature gradient. From this he concluded that these non-local kinetic effects can be very important in the lower transition region and upper chromosphere. Owocki and Canfield[16] using a cruder BGK approximation but more realistic atmospheric model did not find a significant effect in the lower transition region. They did suggest however that non-local heat transport might be important in the upper transition region, depending sensitively on the assumed steepness of the temperature gradient.

The extended FAL models described above have sufficiently steep temperature gradients, where $K \leq 0.1$, that it is not immediately obvious how important the collisionless contribution to the heat flux might be. We have combined the solution of the HVL equation for electrons above two thermal velocities with the SH solution at lower velocities, for these models. The resultant heat flux is shown in figure 3, which also shows the Spitzer-Härm heat flux for comparative purposes. There are no significant differences except in the neighbourhood of the loop apex. There, symmetry constraints prevent the SH approximation from achieving any non-maxwellian solution. Figure 3 shows the results for FAL model C, but similar results were obtained for all the other FAL models. We can conclude therefore that the collisionless electrons do not play any significant role in the energy transport within these quiet solar loops. Neither do they appear to influence the ionization balance within the plasma to any significant degree. This cannot be the explanation for the differential emission measure deficit.

To increase the influence of collisionless electrons requires a more extreme combination of low density and steep temperature gradient than is achieved in these models. The most likely candidate is the solar wind. Weak influence of collisionless electrons on energy transport has been found[17] in the low corona in a model possibly relevant to preflare loops. Flare loops which would not have dramatically steeper temperature gradients in the corona would appear to have densities too large to permit this effect.

AMBIPOLAR DIFFUSION AND NON-EQUILIBRIUM IONIZATION

The ionization balance of any element is described by a continuity equation of the form

$$\frac{\partial N^m}{\partial t} + \boldsymbol{\nabla} \cdot (\boldsymbol{v} N^m) = (\alpha^{m+1} N^{m+1} - \alpha^m N^m) n_e + (S^{m-1} N^{m-1} - S^m N^m) n_e \tag{10}$$

where N^m is the number density and $\boldsymbol{v}$ the fluid velocity for ions in the mth ionization stage, and S and α are the ionization and recombination rates respec-

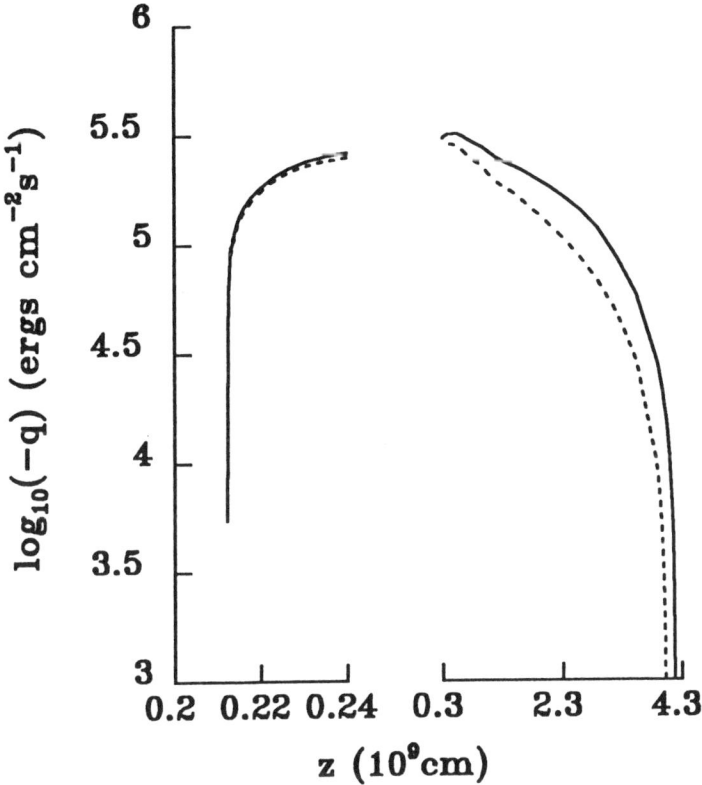

Fig. 3. Heat flux as a function of position z in model C. The solid line is the heat flux q from the calculation of MacNeice et al[11], while the dotted line is the classical result of equation (2).

tively. Equilibrium ionization refers to the case in which the left hand side of this equation is set to zero. Then the ionization balance is determined by the ratio of local ionization and recombination rates.

If flow velocities exist with characteristic timescales

$$\tau \sim \min\left(\frac{1}{|\nabla v|}, \frac{N^m}{|v \cdot \nabla N^m|}\right) \qquad (11)$$

shorter than the recombination or ionization timescales, then the ionization balance will depart from equilibrium. In low density plasmas these rates can be slow and if steep concentration gradients ∇N^m also exist then moderate flow velocities may be sufficient to produce substantial non-equilibrium effects.

This process has been considered before with regard to anomalies in the He I and He II spectra[18], for coronal loop flows[19,20] and in a more general parameter study[21]. We shall now discuss how ambipolar diffusion has been shown to produce non-equilibrium ionization of hydrogen[10].

The FAL models include the diffusion of neutral hydrogen atoms with respect to electrons and protons. This is driven by the steep gradient that exists in the concentration of neutral hydrogen and by the coexisting steep temperature gradient. A nett upward flux of hydrogen is balanced by a nett downward flux of electrons and protons. Thus a diffusively driven cycle develops which has two significant effects on the energy balance. The first is to establish a nett downward flux of ionization energy. In the FAL models this becomes the dominant contribution to the energy flux below 30,000K. The second effect is to alter the spatial distribution of neutral hydrogen by weakening the steep gradient. Neutral hydrogen is found to exist in sufficient concentration to make it the principal source of radiative losses at temperatures as high as 50,000K.

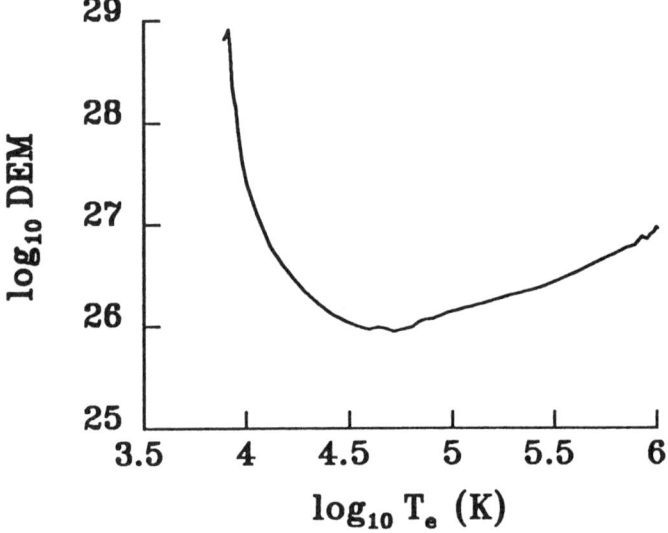

Fig. 4. The differential emission measure as a function of temperature in model C.

The differential emission measure for the FAL model C is shown in figure 4. It has a minimum at $\log_{10} T_e = 4.7$, from which it rises steeply on each side. Although this agrees qualitatively with the differential emission measure curves derived from observations[6] detailed quantitative agreement should not be expected for a number of reasons. The first is that the diffusion of helium should be considered in the energy balance since it is expected to generate substantial energy flow[22]. Secondly diffusion of the heavier elements including helium will alter the ionization balance and thus the radiative losses throughout the models. The ionization balance acquires a non-local character and it is not clear how to take account of this when deriving a differential emission measure from observed line intensities. Clearly direct comparison of computed and observed intensities for individual emission lines is preferable to a comparison of differential emission measure.

REFERENCES

1. L. Spitzer and R. Härm, Phys. Rev. 89, 977 (1953).
2. A.H. Gabriel, Phil. Trans. Roy. Soc. London A. 281, 339 (1976).
3. G.W. Pneuman and R.A. Kopp, Solar Phys. 57, 49 (1978).
4. R.G. Athay, Ap. J. 263, 982 (1982).
5. D.T. Woods, T.E. Holzer and K.B. McGregor, Ap. J. 355, 295 (1990).
6. J.C. Raymond and J.G. Doyle, Ap. J. 247, 686 (1981).
7. D. Rabin and R. Moore, Ap. J. 285, 359 (1984).
8. S.K. Antiochos and G. Noci, Ap. J. 301, 440 (1986).
9. E.C. Shoub, Ap. J. 266, 339 (1983).
10. J.M. Fontenla, E.H. Avrett and R. Loeser, Ap. J. 355, 700 (1990).
11. P. MacNeice, J.M. Fontenla and N.N. Ljepojevic, Ap. J. 369, 544 (1991).
12. E.H. Avrett, in Chromospheric Diagnostics and Modelling, ed. B.W. Lites (Sunspot National Observatory, 1985), p. 67.
13. J.E. Vernazza, E.H. Avrett and R. Loeser, Ap. J. Supp. 45, 635 (1981).
14. L. Landau, Phys. Z. Sowjetunion 10, 154 (1936).
15. D.R. Gray and Kilkenny, Plasma Phys. 22, 81 (1980).
16. S.P. Owocki and R.C. Canfield, Ap. J. 300, 420 (1986).
17. N.N. Ljepojevic and P. MacNeice, Phys. Rev. A 40, 981 (1989).
18. C. Jordan, M.N.R.A.S. 170, 429 (1975).
19. J.T. Mariska, Ap. J. 281, 435 (1984).
20. G. Noci, D. Spadaro, R.A. Zappala and S.K. Antiochos, Ap. J. 338, 1131 (1989).
21. J.A. Joselyn, R.H. Munro and T.E. Holzer, Ap. J. Supp. 40, 793 (1980).
22. T. Nowak and P. Ulmschneider, Astron. and Astrophys. 60, 413 (1977).

MHD TURBULENCE IN AN EXPANDING ATMOSPHERE

M. Velli, R. Grappin and A. Mangeney
Observatoire de Paris-Meudon, 92195 Meudon, France

ABSTRACT

The evolution of MHD fluctuations advected by the solar wind is profoundly affected by the spherical wind expansion: the latter is an important source of anisotropy, the solar plasma becoming increasingly stretched in the transverse direction as it recedes from the sun. As a consequence, the nonlinear evolution of compressive velocity fluctuations in the transverse direction is "frozen" at a finite time in the low frequency limit; at higher frequencies the evolution, e.g, of Alfvén waves is easily predicted only in the linear limit. In the fully three-dimensional and nonlinear case, numerical simulations are necessary: here we present preliminary numerical results from simulations of MHD turbulence in a plasma volume expanding with the wind.

INTRODUCTION

The evolution of MHD fluctuations in the solar wind is of interest both in itself (the solar wind is the only example of a high kinetic and magnetic Reynolds number MHD flow to which we have access in situ) and because of its importance for the average wind dynamics: the simplest theories of the solar wind expansion cannot account for the extremely high velocities often found at solar minimum, which seem to require a supplementary energy input from the waves[1]. In the high velocity solar wind streams the MHD fluctuations (in the frequency range 10^{-4} and 10^{-2} Hz) are dominated by large amplitude quasi-incompressible Alfvén waves propagating away from the sun, and a considerable amount of effort has been devoted to the understanding of their properties and their relation to homogeneous MHD turbulence theory [2,3].

To lowest order, the thermodynamic properties of the average wind and of the fluctuations are governed respectively by the adiabatic law and the WKB law (for Alfvén waves), i.e. specific energies evolve as: $1/R^{4/3}$ and $1/R$. This analysis neglects the energy exchanges which may come from the turbulent dissipation of the waves. Indeed, observations show that the average temperature in the wind often decreases slower than adiabatically[4] and that at least at high frequencies the energy content in the fluctuations decreases more rapidly[5] than $1/R$.

The full self-consistent problem of the combined evolution of the average and fluctuating wind is far from being solved, due to the enormous number of degrees of freedom involved; we concentrate here on a necessary first step: the nonlinear dynamics of the fluctuations in a given global expansion. As just mentioned above, Alfvén wave WKB theory fails in describing the detailed evolution of MHD fluctuations. The origin of the disagreement lies in the fact that the nature of the fluctuations differs from the pure Alfvén states assumed by the theory: there are significant departures both from the kinetic-magnetic energy equipartition and the velocity-magnetic field alignment; the implication is that nonlinear interactions should take place, but the departure from the ideal unidirectional Alfvén waves seems to increase with heliocentric distance[6], contrary to what is predicted by homogeneous MHD turbulence. The expansion of the solar wind might be responsible for this evolution[7], but as we shall see, there are important difficulties with the closure model approach developed so far[8], and we believe that numerical simulations are necessary to advance: we shall present preliminary results in this direction.

SOLAR WIND MHD TURBULENCE

The equations describing magnetic field (b) and incompressible velocity (v) fluctuations, in a plasma of density ρ, may be conveniently expressed in terms of Elsässer variables $z^\pm = v \mp \text{sgn}(B_0) b/\sqrt{(4\pi\rho)}$, which in a homogeneous medium represent two eigenmodes, describing Alfvén waves propagating in opposite directions along the average magnetic field B_0. Because of the time-dependence and inhomogeneity of the backround medium an appropriate average, or separation of scales, must first be carried out and the fluctuation equations are then obtained by subtraction from the original MHD system:

$$\frac{\partial z^\pm}{\partial t} + (U \pm V_a) \cdot \nabla z^\pm + z^\mp \cdot \nabla(U \mp V_a) + \frac{1}{2}(z^- - z^+)(\nabla \cdot V_a \mp \frac{1}{2}\nabla \cdot U)$$
$$= -\frac{1}{\rho}\nabla p^T - (z^\mp \cdot \nabla z^\pm - \langle z^\mp \cdot \nabla z^\pm \rangle), \qquad (1)$$

where p^T is the fluctuation in the total (kinetic and magnetic) pressure, U is the average wind velocity, and V_a the average Alfvén velocity. The first two terms in eqn.(1) describe wave propagation; the third term describes the reflection of waves by the gradients of the equilibrium fields along the fluctuations, and is clearly anisotropic, depending explicitly on the polarisation of the waves. The fourth, isotropic, term describes the WKB losses and the reflection due to the expansion. On the rhs are the nonlinear terms and the total pressure gradient, which for incompressible fields is also nonlinear, and of the same form ($z^\mp \cdot \nabla z^\pm$).

Tu et. al.[9] first successfully described the steepening of an initial flat spectrum of turbulence dominated by outward propagating waves by neglecting the linear coupling of plus and minus modes entirely, while retaining an assumed given fraction of inward propagating waves, in order to keep the nonlinear interactions on. They thus dealt with a closed equation for the (isotropic) energy spectrum E^+. Although the model for the nonlinear term was very crude, the fit of the decay of the energy spectrum with distance was remarkable. The model could not however predict the respective evolution of inward and outward propagating waves.

Zhou and Matthaeus[7] remarked that retaining the nonlinear terms while neglecting the reflection terms (the basic approximation of the Tu et. al. model) is not consistent with the observed levels of $z^\pm$; they therefore began to develop more complex models, based on equations for thr second order moments of eqn.(1), i.e., the $E^\pm$ energy tensors and the real and imaginary parts (symmetric and antisymmetric) of the residual energy tensor $F = z^+z^-$. In general, the equations involve 24 coupled moments, which are then reduced to three by assuming complete (mirror symmetric) isotropy. The latter assumption also destroys the phase relations between inward and outward propagating waves. The equations then become equivalent to the low-frequency limit of eqn.(1) and neglect of nonlinear terms yields an evolution in which the dominance of the outward propagating mode decreases with distance in a way which is qualitatively,(but not quantitatively), consistent with the observations[6].

The isotropy assumption, a necessary step in the derivation of the results just described, is not however easily justified. A relative equipartition between the three components of the magnetic fluctuations in the solar wind is observed, with a maximum power in the plane perpendicular to the average magnetic field. The isotropy assumption discussed above however concerns the energy distribution in wavevector space. There is almost no direct information on this spatial distribution, since satellites measure currently only the variations along the radial direction; in this respect some analyses of in situ data which pretend to show local isotropy[10] are not entirely convincing, as they assume symmetry around the average magnetic field. Actually, the observational data can be interpreted as indicating a very high spatial anisotropy[11], reminiscent of the linear predictions[12] so that it would seem that the possible moderating effects of nonlinear interactions might be limited. Furthermore, there are theoretical reasons to expect a certain amount of anisotropy: for a spherically symmetric configuration, eqn.(1) leads to a different effect of the large scale expansion on fluctuations transverse and parallel to the radial direction. We therefore believe that isotropic models are not capable of a realistic description of solar wind MHD turbulence; to make progress on the fully anisotropic problem, we turn to numerical simulations.

MHD TURBULENCE IN AN EXPANDING BOX

To investigate the effects of the expansion, we consider the evolution of a plasma parcel moving in a simplified but realistic model of the wind; because we are interested in the evolution in the supersonic and superalfvénic regions, we take the wind speed U to be constant, and the expansion to be spherical. If the initial cross-section of the plasma volume a_0 (which we will also take to be the initial thickness of the plasma parcel) is much smaller than the initial distance from the sun, R_0, i.e. $\epsilon = a_0/R_0 \ll 1$, we may neglect the curvature effects as well as the variation of the cross section of the box in the radial direction at any given time (see fig. 1).

We are left with a rectangular box, whose cross section a (in the xy plane) evolves as $a = a_0 + U\epsilon t$, but whose thickness (in the z direction) remains constant. Conservation of mass implies that the average density, $\bar{\rho}$, decreases as a^{-2}, and it is convenient to extract this dependence from the density, pressure and magnetic field, i.e. to write the equations in terms of $\tilde{\rho} = \rho/\bar{\rho}, \tilde{p} = p/\bar{\rho}, \tilde{b} = b/\sqrt{\bar{\rho}}$. We will place ourselves in the comoving frame, by introducing the "coexpanding" coordinates $\tilde{x} = x/a$, $\tilde{y} = y/a$, $\tilde{z} = z - Ut$ and the gradient operator $\nabla = (\partial/a\partial\tilde{x}, \partial/a\partial\tilde{y}, \partial/\partial\tilde{z})$; in this frame the residual velocity variable is defined by $u = v - U$ where $U = (\epsilon U x, \epsilon U y, U)$. Dropping the tilde from the renormalized variables, the equations of motion and magnetic induction become

$$\frac{\partial u}{\partial t} + u \cdot \nabla u + \frac{\dot{a}}{a}\overline{\overline{P}}_\perp \cdot u = -\frac{1}{\rho}\left(\nabla\left(p + \frac{b^2}{8\pi}\right) + \frac{b \cdot \nabla b}{4\pi}\right) \qquad (2)$$

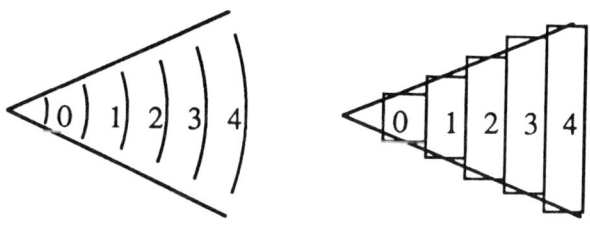

Fig. 1. Approximation of a spherically expanding plasma packet by a rectangular box.

$$\frac{\partial \boldsymbol{b}}{\partial t} - \frac{\dot{a}}{a}\overline{\overline{\boldsymbol{P}}}_\parallel \cdot \boldsymbol{b} = \nabla \times (\boldsymbol{u} \times \boldsymbol{b}), \qquad (3)$$

here $\overline{\overline{\boldsymbol{P}}}_\perp$ and $\overline{\overline{\boldsymbol{P}}}_\parallel$ are the projectors on the xy plane and on the z direction respectively. We don't write the equations of continuity and energy conservation: the former retains its standard form in the normalized variables, while the latter, written in terms of the pressure p, contains an additional expansion cooling term $\dot{a}/a(\gamma-1)p$ (γ is the ratio of specific heats). From eqns. (2) and (3) it is clear that the expansion acts differently on the velocity and magnetic field. The transverse components of u feel a drag, while it is the parallel component of the magnetic field which is damped (this expresses the conservation of magnetic flux through the box). A particularly important effect of the drag[13] which we want to stress in this communication is the inhibition of nonlinear interactions, as may be seen by considering a simple one dimensional model neglecting magnetic field and pressure. The x or y component of the equation of motion then becomes a Burgers equation with damping,

$$\frac{\partial u}{\partial t} + \frac{u}{a}\frac{\partial u}{\partial x} + \frac{\dot{a}}{a}u = 0, \qquad (4)$$

which may reduced to the standard Burgers equation by the transformation of variables

$$\tilde{u} = a/a_0 u, \qquad \theta = \int_0^t d\tau a_0^2/a^2 = R_0/U(1 - 1/(1 + Ut/R_0)). \qquad (5)$$

The asymptotic evolution ($t \to \infty$) is frozen in the new variables at the time $\theta = R_0/U$, due to the stretching of the box, which is in competition with the wave steepening. Depending on the ratio of this time to shock formation time, the frozen state will be more or less developed. A 2-d simulation of an Orszag-Tang vortex (one direction expanding, one stationary), leads to a similar freezing (fig. 2). The presence of a magnetic field of course makes this simple picture incomplete, but these results show how the effect of the nonlinear interactions may be altered by the solar wind expansion. In particular, nonlinear effects are unable to prevent the growth of a large-scale anisotropy.

158 MHD Turbulence in an Expanding Atmosphere

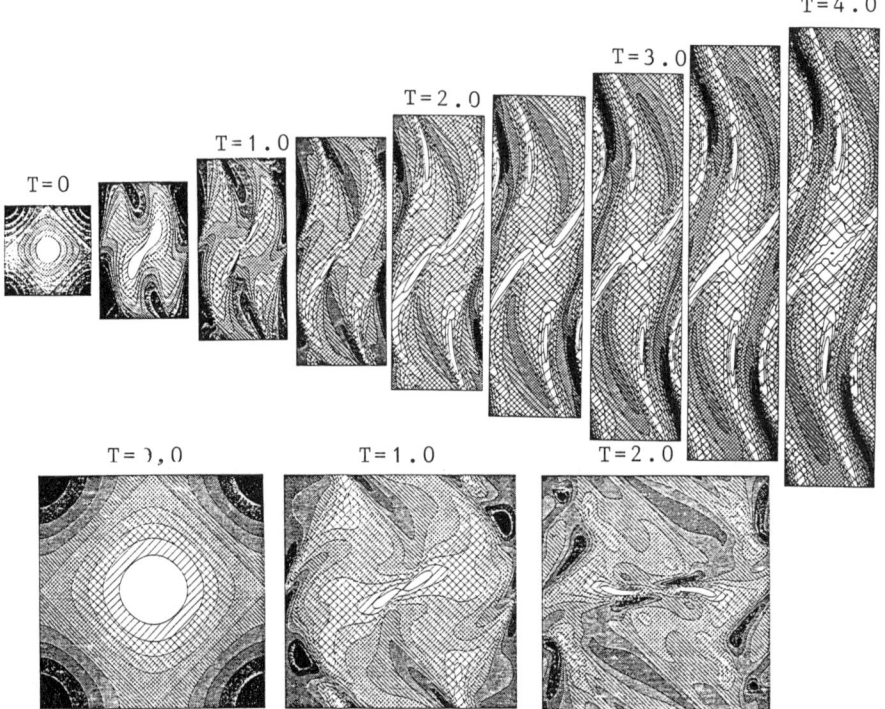

Fig. 2. 2-D simulation of an Orszag-Tang vortex. Top, expanding box. Bottom, standard evolution. The top evolution is effectively frozen at T=1.0.

REFERENCES

1. Leer E., T.E., Holzer, and T. Fla, Space Sci. Rev. 33, 161 ((1982)).
2. Mangeney, A., Grappin, R. and Velli, M., in Advances in Solar System MHD, E.R. Priest, A.W. Hood eds. (Cambridge, 1991), p. 327.
3. Marsch, E., in Physics of the inner heliosphere, R. Schwenn and E. Marsch eds. (Springer-Verlag, 1991).
4. Freeman, J. W., Geophys. Res. Letts. 15, 88 (1988).
5. Bavassano, B., and E. J. Smith, J. Geophys. Res. 91, 1706 (1986).
6. Roberts, D. A., M. L. Goldstein, L. W. Klein, and W. H. Matthaeus, J. Geophys. Res. 92, 12023 (1987).
7. Zhou Y. and W. H. Matthaeus, J. Geophys. Res. 95A, 10291 (1990).
8. Velli, M., R. Grappin, and A. Mangeney, Geophys. Astrophys. Fluid Dyn. , (1991, in press).
9. Tu, C.Y., Z. Y. Pu, F. S. Wei, J. Geophys. Res. 89, 9695 (1984).
10. Matthaeus W.H. , M.L. Goldstein and D. A. Roberts, J. Geophys. Res. 95, 20673 (1990).
11. Carbone V., P. Veltri, F. Malara, preprint , (1991).
12. Völk H.J. and W. Alpers, Astrophys. Space Sci. 20, 267 (1973).

13. U. Frisch, private communication

HOT MASS TRANSPORT IN THE SOLAR ACTIVE PROMINENCE

A. Kučera
Astron. Institute, Slovak Acad. of Sciences, Tatranská Lomnica, Czech.

M. Saniga
Astron. Institute, Slovak Acad. of Sciences, Tatranská Lomnica, Czech.

J. Rybák
Astron. Institute, Slovak Acad. of Sciences, Tatranská Lomnica, Czech.

ABSTRACT

On October 16, 1990, a remarkable active prominence was observed, in Hα line, during 113 minutes. The time gap between successive profile recordings was 20 seconds and 339 profiles have been recorded altogether. We focused our attention on Doppler shift measurements as well as on significant changes in the prominence intensity. We also give a possible explanations of a sudden increment in the intensity of Hα line in dependence on the fact if the whole profile originates in one particular place in the prominence or if it is a superposition of the radiation incident from two different, physically unrelated parts of the prominence.

INTRODUCTION

Prominences are well suited for investigating the energy balance, radiative transfer, and complex interchange of mass between the solar photosphere, chromosphere and corona. They are still a subject of interest in solar physics. In recent years, there has been an important increase in the quantity and quality of ground-based observations of prominences, especially those using spectrophotometrical methods. The spectral analysis and morphological properties of prominences usually provide us with important information on the dynamics, radiative transfer, and stability of the prominences, for example see [2,7,8].

Of a great interest are the prominences with central reversal in their Hα profile since they may display appreciable variations in activity. Such observations have also been included in our scientific programme with the Stará lesná spectrograph, figure 1, because a long series of Hα profiles of prominences can be recorded. A mass and energy transport in magnetic structures is possible to study from such an observational material.

OBSERVATIONS

Activity in a previously quiescent prominence was seen to develop on the west limb (position angle 242^0) on October 16, 1990, and was recorded between 10:45 and 12:39 UT. The prominence was situated in the active regions NOAA 6306 and 6302. A rectangular slit area, 2" x 0.25", was aimed at location of the most interesting Hα emission profile with central reversal, near the centre of the prominence, figure 2. A total number of 339 profiles of the Hα emission from

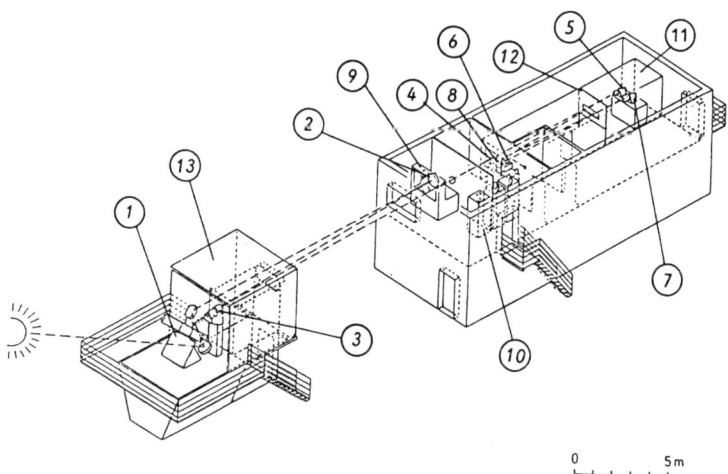

Fig. 1. Optical path of light beam and the sketch of the instrument. (1) coelostat, (2) primary mirror, (3) secondary mirror of the telescope, (4) entrance slit of the spectrograph, (5) collimator, (6) diffraction grating, (7) camera, (8) focal planes of camera, (9) guider, (10) controlling electronic device, (11) tunnel, (12) blackened walls protecting against the parasitic light, (13) moveable roof of the coelostat.

this location was detected with a time delay of about 20 seconds. Each profile corresponds to 11 Å with Hα line centered in the first order.

Observations were carried out at the Stará Lesná Observatory (Czechoslovakia) with horizontal solar telescope equipped with a spectrograph. The telescope of an off-axial Kutter design (D=0.5 m, f=35 m), illuminated through a coelostat, creates the Sun's image of 32.5 cm in a diameter (1 mm=5.9'') with theoretical resolution 0.25'' (0.04 mm).

The spectrograph of the Czerny-Turner type (f=9.65 m for both the collimator and camera mirrors), equipped with a Bausch & Lomb diffraction grating (204x156 mm, 632 lines per mm, blaze angle 26^0) gives a dispersion of 1.62 Å mm^{-1} for the wavelength 6500 Å (the first order). The resolution achieved is about 90 000 (i.e., 3/4 of the theoretical value). More detail information about the instrument are in papers [1,3].

An image tube consisting of a one-stage image intensifier and a vidicon electronic camera (TV Nightvision, Vilati, Hungary) was used as the spectrum detector. With the aid of an additional electronic device we then selected a "spectral scan" of 2'' x 11 Å from the whole spectral image, and subsequently digitized it into 480 data (each containing 8-bit information); "a data distance" was 0.015 mm that coresponds to the dispersion 2.34 mÅ in the first order. Since a sufficient S/N ratio was achieved, an integration time of 0.64 seconds was chosen. The necessary dark current measurements as well as the Hα profiles of the quiet portion of the disk centre were also recorded.

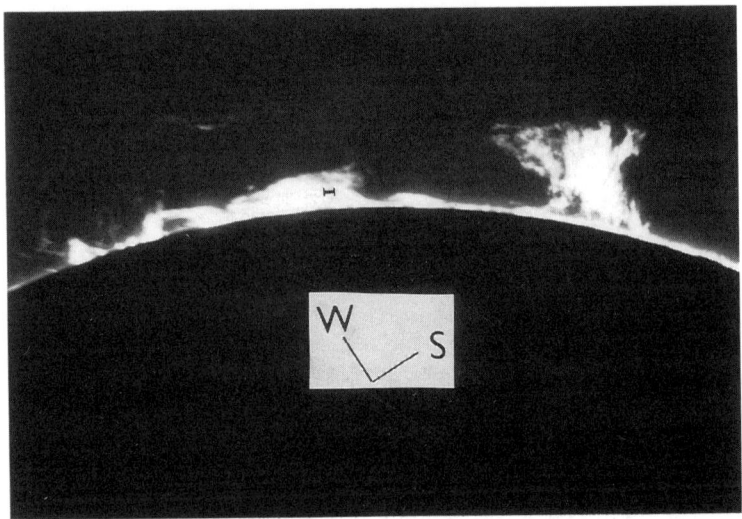

Fig. 2. The detail of the investigated prominence with entrance slit position indicated.

During the whole observation an entrance slit jaw video movie of the actual slit position was also recorded using a Day Star Hα filter. Seeing was, on the average, about $3-4''$.

DATA REDUCTION

Using the video movie of the slit jaw, the profiles were selected very carefully; we finally considered only those of them which were not shifted by seeing. The dark current was substracted from all the emission data, and a "flat scan" correction for unequal sensitivity of the detector's rows was also taken into account.

The Hα profiles of the disk centre were used for wavelength calibrations of the emission measurements. In order to determine the absolute intensity of the emmision profiles, the transmissivity of the neutral density filter as well as the absolute intensity data of the disk centre, acording [5,6] were also considered. An approximative reduction of the scattered atmospheric light (from disk to aureola) was carried out; corrections for a differential extinction of the Earth's atmosphere were not applied since the observational time was around noon.

DISCUSSION

The active phase of the enhanced emission in the Hα line lasted for 24 minutes; selected profiles representing this phase are shown in figures 3 and 4. Later, remarkable changes in the shape of prominence were seen 45 minutes after this enhancement in the Hα line intensity, as illustrated in figure 5. There are several theoretical explanations of these events:

(1) The whole Hα emission profile originates in one particular place in

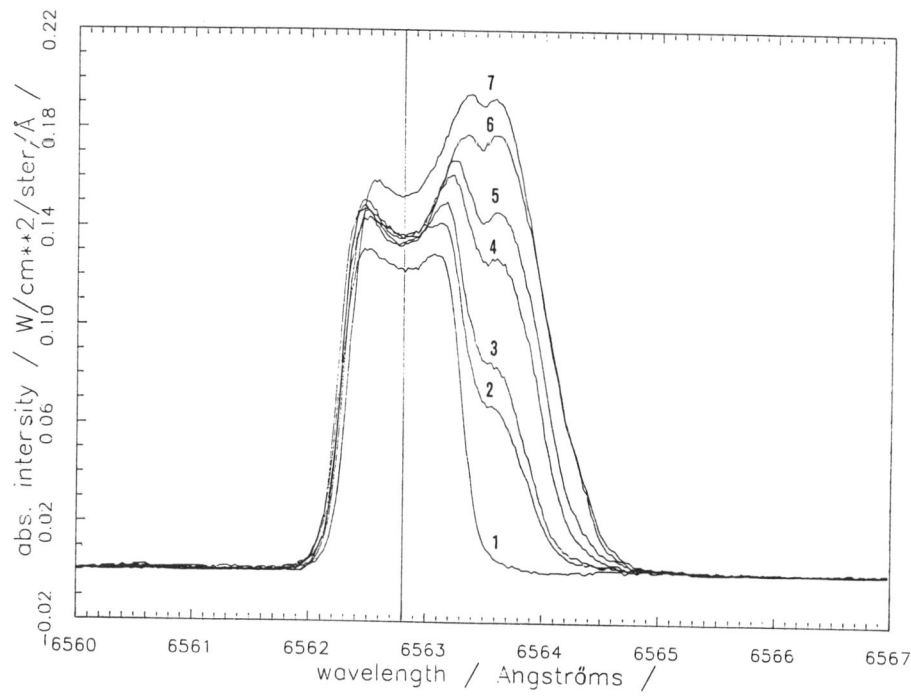

Fig. 3. The increasing phase of the Hα emission evolution. (profile number / time[UT]) : 1/10:52:20, 2/10:55:08, 3/10:55:29, 4/10:56:15, 5/10:56:23, 6/11:00:02, 7/11:00:44.

the prominence. The red Doppler shift, approximately 0.46Å, implies a line-of-sight velocity of approximately 21 km sec^{-1}. A peculiar shape of the profile should then be the result of an inhomogeneous temperature of the moving part of prominence (different temperatures and their gradients). A FWHM increment from the value 1.06Å to 1.74Å as well as the increase in the Hα line intensity evidence a rapid increase in the temperature. As the resulting profiles are rather complicated, a more detailed treatment of this situation (non-LTE calculations) is necessary. Here it is worth noting that the tangential velocity of the front of the prominence's "moving tongue" is roughly 20 km sec^{-1}, too (see figure 5).

(2) The Hα emission profile is a superposition of the radiation from two (or more) different, physically UNRELATED parts of the prominence. In order to discuss this possibility we first made a substraction of the profile no.1, which is supposed to reflect the physical situation in a relatively stable and non-movig part of prominence from all the other ones. The results of substraction are

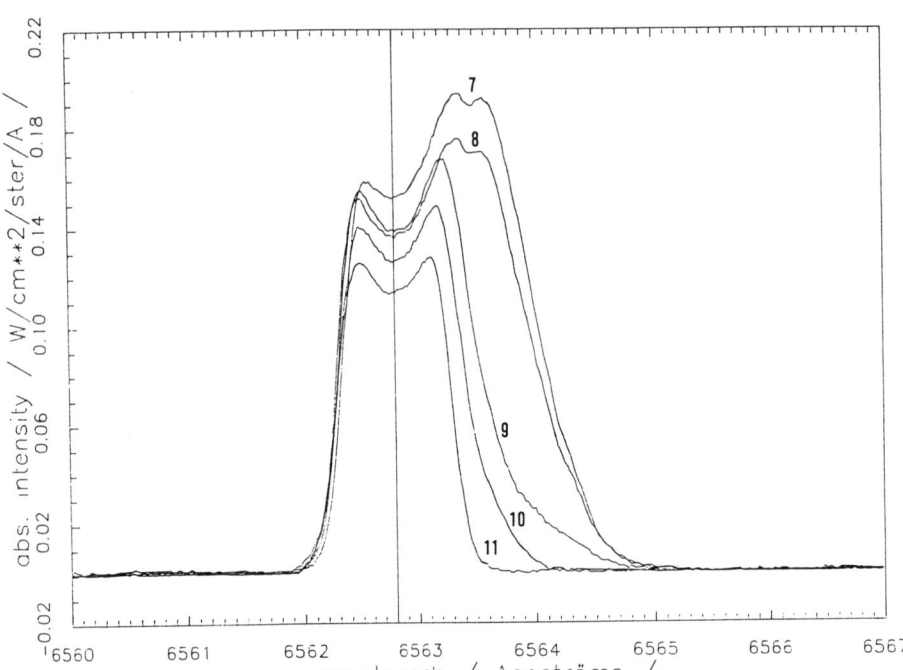

Fig. 4. The decreasing phase of the Hα emission time evolution. (profile number / time[UT]) : 7/11:00:44, 8/11:01:26, 9/11:06:20, 10/11:10:53, 11/11:16:08.

shown in figures 6 and 7; where "I" denotes "2-1" substraction, "II" - "3-1", etc. Several interesting findings can be drawn !

The INCREASING PHASE is characterized by a considerable red shift of the maximum of the emission part of profile, at about 0.77Å, which corresponds to line-of-sight velocity of 35 km sec^{-1}. As the increment of the intensity of Hα line is not accompanied by the change in wavelenght shift we arrive at the conclusion that a hot element of prominence plasma is moving steadily, changing, however, its temperature and/or total mass density along the line-of-sigth.

During the DECREASING PHASE we observe both a decrement in intensity and a shift of emission maxima toward "laboratory" center of Hα line. It would correspond to the fact that the moving element is slowed-down and its temperature is falling also. However, the observed phenomenon can also originate artificially, if it not possible to keep a slit pointing at one and the same place of the prominence's image during the time of observation as shown[4] (profiles 6

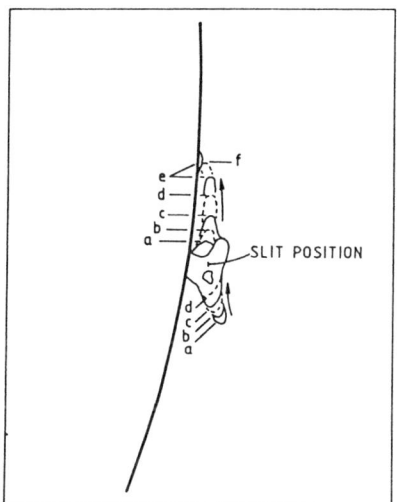

Fig. 5. The time evolution [UT] of the prominence during the observation. a/11:59:41, b/12:01:47, c/12:04:14, d/12:05:38, e/12:06:41, f/12:08:05.

and 7 in their figure 5).

In both increasing and decreasing phases of activity, we can see a complicated shape of "substracted" profiles near the line center and in the blue wing; it seems that the intensity decrement in the blue wing and the enhanced emission in the red wing are significantly correlated. In almost all cases a slight trace of the "central reversal" effect can be noticed.This, when taking into account a "blue increment", means that the two differently moving parts of prominence are not totally unrelated, but here should exist a fine coupling between them.

To understand the origin of this coupling (dynamics, radiative transfer) a detailed non-LTE modelling of the physical situation is necessary. These calculations are being in the progress now.

In conclusion we would like to stress that the time variations of the Hα profiles as well as the subsequent morphological changes of the prominence, as described above, not only reflect the dynamics (i.e. motion of prominence plasma) but they may also well be related to the energy release in the prominence. However, only detailed calculations (radiative cooling included) can show if the energy release was situated inside, behind or in front of the prominence and what was an initializing event of such enegy release and mass transfer.

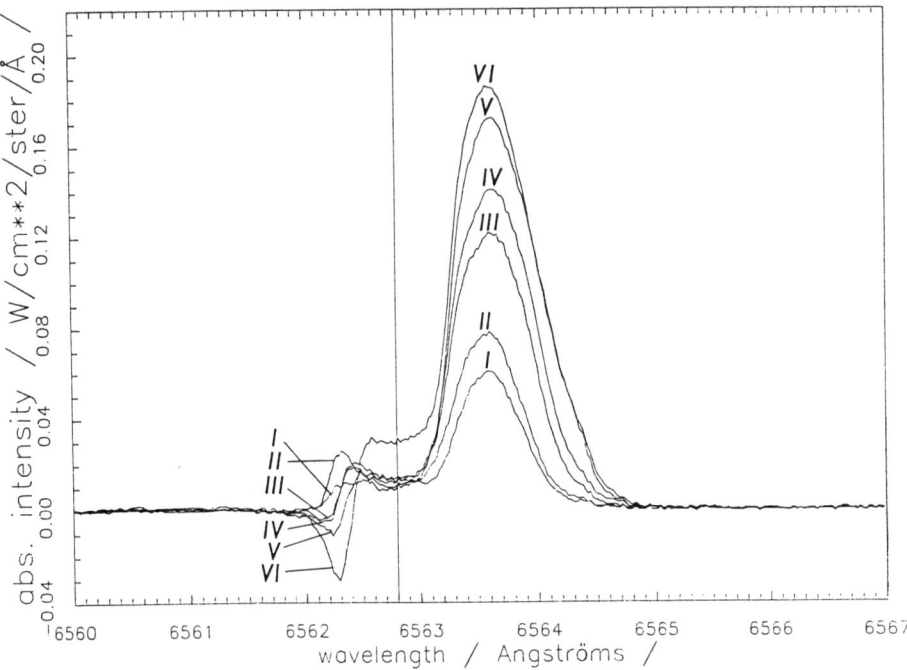

Fig. 6. Difference profiles of the increasing phase, where the first profile from figure 3 is substracted from the remaining ones. (I=(2-1),II=(3-1), III=(4-1),IV=(5-1),V=(6-1),VI=(7-1))

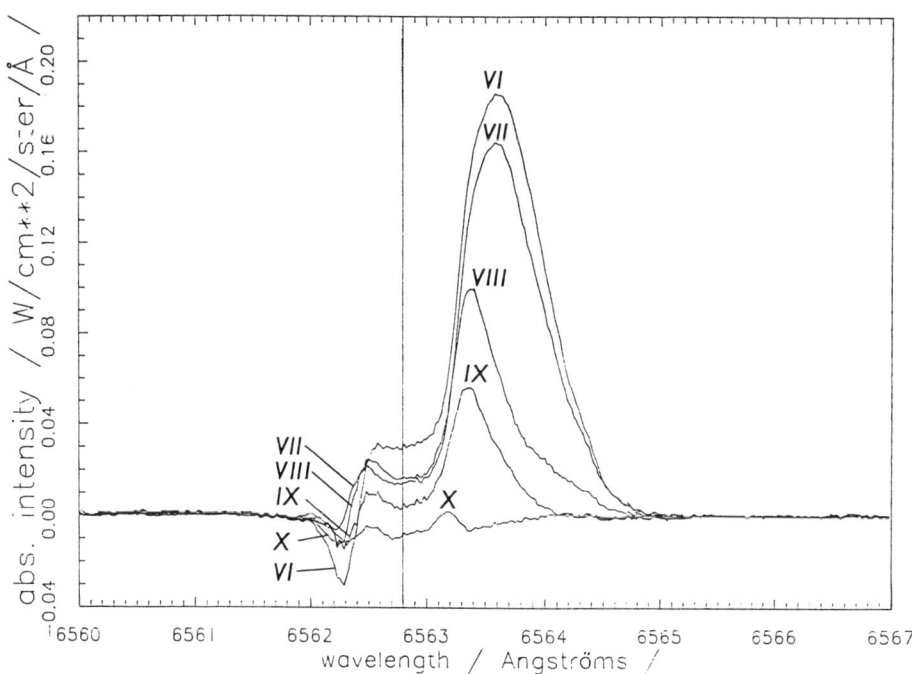

Fig. 7. The same as in figure 6 for the decreasing phase. (VI=(7-1), VII=(8-1),VIII=(9-1),IX=(10-1),X=(11-1)).

REFERENCES

1. P. Ambrož, V. Bumba, M. Klvaňa and P. Macák, Physics Solaire 14, 107 (1980).
2. P. Lemaire, D. Samain, and J.C. Vial, Adv. Space. Res. 10, 191 (1990).
3. A. Kučera, J. Rybák, M. Minarovjech, D. Novocký and M. Saniga, Astroph. Space Sci. 17, 281 (1990).
4. J.T. Mariska, G. A. Doschek and V. Feldman, Ap. J. 232, 929 (1979).
5. H. Neckel and D. Labs, Solar Phys. 90, 205 (1984).
6. H. Neckel and D. Labs, Solar Phys. 95, 229 (1985).
7. B. Vršnak, V. Ruždjak, R. Brajša and F. Zloch, Solar Phys. 127, 129 (1990).
8. J.B. Zirker and S. Koutchmy, Solar Phys. 127, 109 (1990).

ACKNOWLEDGEMENTS

This work has been supported under a Grant GA 494/1991 by the Slovak Academy of Sciences.

Author Index

A
Angrilli, F., 126
Antonucci, E., 126

B
Bruner, M. E., 126
Bogdan, T. J., 1

C
Ciminiera, L., 126
Cook, J. W., 55

D
de Assis, A. S., 121
de Azevedo, C. A., 121
Dere, K. P., 63
Dodero, M. A., 126

E
Evans, B. L., 126

F
Finn, J. M., 71, 79
Fossi, B. M., 126

G
Galsgaard, K., 13
Golub, L., 126, 136
Goode, P. R., 85
Grappin, R., 154

K
Kučera, A., 160

L
Landini, M., 126
Lau, Y.-T., 71
Longcope, D. W., 100
Lou, Y.-Q., 116

M
MacNeice, P., 145
Malvezzi, M., 126
Mangeney, A., 154
Martens, P. C. H., 111
McWhirter, P., 145

N
Neidig, D. F., 126
Noci, G., 126
Nordlund, Å., 13

P
Peres, G., 136, 140
Perona, G., 126
Poletto, G., 126

R
Reale, F., 136, 140
Roberts, B., 24
Rybák, J., 160

S

Sakanaka, P. H., 121
Saniga, M., 160
Schmidt, W. K. H., 126
Seehafer, N., 35
Shigueoka, H., 121
Stenflo, J. O., 40
Sudan, R. N., 100
Sun, M. T., 111

T

Thomas, R. J., 126
Tondello, G., 126

V

Velli, M., 154

W

Wu, S. T., 111

AIP Conference Proceedings

		L.C. Number	ISBN
No. 1	Feedback and Dynamic Control of Plasmas – 1970	70-141596	0-88318-100-2
No. 2	Particles and Fields – 1971 (Rochester, NY)	71-184662	0-88318-101-0
No. 3	Thermal Expansion – 1971 (Corning, NY)	72-76970	0-88318-102-9
No. 4	Superconductivity in d- and f-Band Metals (Rochester, NY, 1971)	74-18879	0-88318-103-7
No. 5	Magnetism and Magnetic Materials – 1971 (2 parts) (Chicago, IL)	59-2468	0-88318-104-5
No. 6	Particle Physics (Irvine, CA, 1971)	72-81239	0-88318-105-3
No. 7	Exploring the History of Nuclear Physics – 1972	72-81883	0-88318-106-1
No. 8	Experimental Meson Spectroscopy – 1972	72-88226	0-88318-107-X
No. 9	Cyclotrons – 1972 (Vancouver)	72-92798	0-88318-108-8
No. 10	Magnetism and Magnetic Materials –1972	72-623469	0-88318-109-6
No. 11	Transport Phenomena – 1973 (Brown University Conference)	73-80682	0-88318-110-X
No. 12	Experiments on High Energy Particle Collisions – 1973 (Vanderbilt Conference)	73-81705	0-88318-111-8
No. 13	π-π Scattering – 1973 (Tallahassee Conference)	73-81704	0-88318-112-6
No. 14	Particles and Fields – 1973 (APS/DPF Berkeley)	73-91923	0-88318-113-4
No. 15	High Energy Collisions – 1973 (Stony Brook, NY)	73-92324	0-88318-114-2
No. 16	Causality and Physical Theories (Wayne State University, 1973)	73-93420	0-88318-115-0
No. 17	Thermal Expansion – 1973 (Lake of the Ozarks)	73-94415	0-88318-116-9
No. 18	Magnetism and Magnetic Materials – 1973 (2 parts) (Boston, MA)	59-2468	0-88318-117-7
No. 19	Physics and the Energy Problem – 1974 (APS, Chicago, IL)	73-94416	0-88318-118-5
No. 20	Tetrahedrally Bonded Amorphous Semiconductors (Yorktown Heights, NY, 1974)	74-80145	0-88318-119-3
No. 21	Experimental Meson Spectroscopy – 1974 (Boston, MA)	74-82628	0-88318-120-7
No. 22	Neutrinos – 1974 (Philadelphia, PA)	74-82413	0-88318-121-5

No. 23	Particles and Fields – 1974 (APS/DPF, Williamsburg, VA)	74-27575	0-88318-122-3
No. 24	Magnetism and Magnetic Materials – 1974 (20th Annual Conference, San Francisco, CA)	75-2647	0-88318-123-1
No. 25	Efficient Use of Energy (The APS Studies on the Technical Aspects of the More Efficient Use of Energy)	75-18227	0-88318-124-X
No. 26	High-Energy Physics and Nuclear Structure – 1975 (Santa Fe and Los Alamos, NM)	75-26411	0-88318-125-8
No. 27	Topics in Statistical Mechanics and Biophysics: A Memorial to Julius L. Jackson (Wayne State University, 1975)	75-36309	0-88318-126-6
No. 28	Physics and Our World: A Symposium in Honor of Victor F. Weisskopf (M.I.T., 1974)	76-7207	0-88318-127-4
No. 29	Magnetism and Magnetic Materials – 1975 (21st Annual Conference, Philadelphia, PA)	76-10931	0-88318-128-2
No. 30	Particle Searches and Discoveries – 1976 (Vanderbilt Conference)	76-19949	0-88318-129-0
No. 31	Structure and Excitations of Amorphous Solids (Williamsburg, VA, 1976)	76-22279	0-88318-130-4
No. 32	Materials Technology – 1976 (APS New York Meeting)	76-27967	0-88318-131-2
No. 33	Meson-Nuclear Physics – 1976 (Carnegie-Mellon Conference)	76-26811	0-88318-132-0
No. 34	Magnetism and Magnetic Materials – 1976 (Joint MMM-Intermag Conference, Pittsburgh, PA)	76-47106	0-88318-133-9
No. 35	High Energy Physics with Polarized Beams and Targets (Argonne, IL, 1976)	76-50181	0-88318-134-7
No. 36	Momentum Wave Functions – 1976 (Indiana University)	77-82145	0-88318-135-5
No. 37	Weak Interaction Physics – 1977 (Indiana University)	77-83344	0-88318-136-3
No. 38	Workshop on New Directions in Mössbauer Spectroscopy (Argonne, IL, 1977)	77-90635	0-88318-137-1
No. 39	Physics Careers, Employment and Education (Penn State, 1977)	77-94053	0-88318-138-X
No. 40	Electrical Transport and Optical Properties of Inhomogeneous Media (Ohio State University, 1977)	78-54319	0-88318-139-8
No. 41	Nucleon-Nucleon Interactions – 1977 (Vancouver)	78-54249	0-88318-140-1

No. 42	Higher Energy Polarized Proton Beams (Ann Arbor, MI, 1977)	78-55682	0-88318-141-X
No. 43	Particles and Fields – 1977 (APS/DPF, Argonne, IL)	78-55683	0-88318-142-8
No. 44	Future Trends in Superconductive Electronics (Charlottesville, 1978)	77-9240	0-88318-143-6
No. 45	New Results in High Energy Physics – 1978 (Vanderbilt Conference)	78-67196	0-88318-144-4
No. 46	Topics in Nonlinear Dynamics (La Jolla Institute)	78-57870	0-88318-145-2
No. 47	Clustering Aspects of Nuclear Structure and Nuclear Reactions (Winnipeg, 1978)	78-64942	0-88318-146-0
No. 48	Current Trends in the Theory of Fields (Tallahassee, FL, 1978)	78-72948	0-88318-147-9
No. 49	Cosmic Rays and Particle Physics – 1978 (Bartol Conference)	79-50489	0-88318-148-7
No. 50	Laser-Solid Interactions and Laser Processing – 1978 (Boston, MA)	79-51564	0-88318-149-5
No. 51	High Energy Physics with Polarized Beams and Polarized Targets (Argonne, IL, 1978)	79-64565	0-88318-150-9
No. 52	Long-Distance Neutrino Detection – 1978 (C. L. Cowan Memorial Symposium)	79-52078	0-88318-151-7
No. 53	Modulated Structures – 1979 (Kailua Kona, Hawaii)	79-53846	0-88318-152-5
No. 54	Meson-Nuclear Physics – 1979 (Houston, TX)	79-53978	0-88318-153-3
No. 55	Quantum Chromodynamics (La Jolla, CA, 1978)	79-54969	0-88318-154-1
No. 56	Particle Acceleration Mechanisms in Astrophysics (La Jolla, CA, 1979)	79-55844	0-88318-155-X
No. 57	Nonlinear Dynamics and the Beam-Beam Interaction (Brookhaven, NY, 1979)	79-57341	0-88318-156-8
No. 58	Inhomogeneous Superconductors – 1979 (Berkeley Springs, WV)	79-57620	0-88318-157-6
No. 59	Particles and Fields – 1979 (APS/DPF Montreal)	80-66631	0-88318-158-4
No. 60	History of the ZGS (Argonne, IL, 1979)	80-67694	0-88318-159-2
No. 61	Aspects of the Kinetics and Dynamics of Surface Reactions (La Jolla Institute, 1979)	80-68004	0-88318-160-6
No. 62	High Energy e^+e^- Interactions (Vanderbilt, 1980)	80-53377	0-88318-161-4
No. 63	Supernovae Spectra (La Jolla, CA, 1980)	80-70019	0-88318-162-2

No. 64	Laboratory EXAFS Facilities – 1980 (Univ. of Washington)	80-70579	0-88318-163-0
No. 65	Optics in Four Dimensions – 1980 (ICO, Ensenada)	80-70771	0-88318-164-9
No. 66	Physics in the Automotive Industry – 1980 (APS/AAPT Topical Conference)	80-70987	0-88318-165-7
No. 67	Experimental Meson Spectroscopy – 1980 (Sixth International Conference, Brookhaven, NY)	80-71123	0-88318-166-5
No. 68	High Energy Physics – 1980 (XX International Conference, Madison, WI)	81-65032	0-88318-167-3
No. 69	Polarization Phenomena in Nuclear Physics – 1980 (Fifth International Symposium, Santa Fe, NM)	81-65107	0-88318-168-1
No. 70	Chemistry and Physics of Coal Utilization – 1980 (APS, Morgantown)	81-65106	0-88318-169-X
No. 71	Group Theory and its Applications in Physics – 1980 (Latin American School of Physics, Mexico City)	81-66132	0-88318-170-3
No. 72	Weak Interactions as a Probe of Unification (Virginia Polytechnic Institute – 1980)	81-67184	0-88318-171-1
No. 73	Tetrahedrally Bonded Amorphous Semiconductors (Carefree, AZ, 1981)	81-67419	0-88318-172-X
No. 74	Perturbative Quantum Chromodynamics (Tallahassee, FL, 1981)	81-70372	0-88318-173-8
No. 75	Low Energy X-Ray Diagnostics – 1981 (Monterey, CA)	81-69841	0-88318-174-6
No. 76	Nonlinear Properties of Internal Waves (La Jolla Institute, 1981)	81-71062	0-88318-175-4
No. 77	Gamma Ray Transients and Related Astrophysical Phenomena (La Jolla Institute, 1981)	81-71543	0-88318-176-2
No. 78	Shock Waves in Condensed Matter – 1981 (Menlo Park, NJ)	82-70014	0-88318-177-0
No. 79	Pion Production and Absorption in Nuclei – 1981 (Indiana University Cyclotron Facility)	82-70678	0-88318-178-9
No. 80	Polarized Proton Ion Sources (Ann Arbor, MI, 1981)	82-71025	0-88318-179-7
No. 81	Particles and Fields – 1981: Testing the Standard Model (APS/DPF, Santa Cruz, CA)	82-71156	0-88318-180-0
No. 82	Interpretation of Climate and Photochemical Models, Ozone and Temperature Measurements (La Jolla Institute, 1981)	82-71345	0-88318-181-9
No. 83	The Galactic Center (Cal. Inst. of Tech., 1982)	82-71635	0-88318-182-7

No. 84	Physics in the Steel Industry (APS/AISI, Lehigh University, 1981)	82-72033	0-88318-183-5
No. 85	Proton-Antiproton Collider Physics – 1981 (Madison, WI)	82-72141	0-88318-184-3
No. 86	Momentum Wave Functions – 1982 (Adelaide, Australia)	82-72375	0-88318-185-1
No. 87	Physics of High Energy Particle Accelerators (Fermilab Summer School, 1981)	82-72421	0-88318-186-X
No. 88	Mathematical Methods in Hydrodynamics and Integrability in Dynamical Systems (La Jolla Institute, 1981)	82-72462	0-88318-187-8
No. 89	Neutron Scattering – 1981 (Argonne National Laboratory)	82-73094	0-88318-188-6
No. 90	Laser Techniques for Extreme Ultraviolet Spectroscopy (Boulder, CO, 1982)	82-73205	0-88318-189-4
No. 91	Laser Acceleration of Particles (Los Alamos, NM, 1982)	82-73361	0-88318-190-8
No. 92	The State of Particle Accelerators and High Energy Physics (Fermilab, 1981)	82-73861	0-88318-191-6
No. 93	Novel Results in Particle Physics (Vanderbilt, 1982)	82-73954	0-88318-192-4
No. 94	X-Ray and Atomic Inner-Shell Physics – 1982 (International Conference, U. of Oregon)	82-74075	0-88318-193-2
No. 95	High Energy Spin Physics – 1982 (Brookhaven National Laboratory)	83-70154	0-88318-194-0
No. 96	Science Underground (Los Alamos, NM, 1982)	83-70377	0-88318-195-9
No. 97	The Interaction Between Medium Energy Nucleons in Nuclei – 1982 (Indiana University)	83-70649	0-88318-196-7
No. 98	Particles and Fields – 1982 (APS/DPF University of Maryland)	83-70807	0-88318-197-5
No. 99	Neutrino Mass and Gauge Structure of Weak Interactions (Telemark, 1982)	83-71072	0-88318-198-3
No. 100	Excimer Lasers – 1983 (OSA, Lake Tahoe, NV)	83-71437	0-88318-199-1
No. 101	Positron-Electron Pairs in Astrophysics (Goddard Space Flight Center, 1983)	83-71926	0-88318-200-9
No. 102	Intense Medium Energy Sources of Strangeness (UC-Santa Cruz, CA, 1983)	83-72261	0-88318-201-7

No. 103	Quantum Fluids and Solids – 1983 (Sanibel Island, FL)	83-72440	0-88318-202-5
No. 104	Physics, Technology and the Nuclear Arms Race (APS, Baltimore, MD, 1983)	83-72533	0-88318-203-3
No. 105	Physics of High Energy Particle Accelerators (SLAC Summer School, 1982)	83-72986	0-88318-304-8
No. 106	Predictability of Fluid Motions (La Jolla Institute, 1983)	83-73641	0-88318-305-6
No. 107	Physics and Chemistry of Porous Media (Schlumberger-Doll Research, 1983)	83-73640	0-88318-306-4
No. 108	The Time Projection Chamber (TRIUMF, Vancouver, 1983)	83-83445	0-88318-307-2
No. 109	Random Walks and Their Applications in the Physical and Biological Sciences (NBS/La Jolla Institute, 1982)	84-70208	0-88318-308-0
No. 110	Hadron Substructure in Nuclear Physics (Indiana University, 1983)	84-70165	0-88318-309-9
No. 111	Production and Neutralization of Negative Ions and Beams (3rd Int'l Symposium) (Brookhaven, NY, 1983)	84-70379	0-88318-310-2
No. 112	Particles and Fields – 1983 (APS/DPF, Blacksburg, VA)	84-70378	0-88318-311-0
No. 113	Experimental Meson Spectroscopy – 1983 (7th International Conference, Brookhaven, NY)	84-70910	0-88318-312-9
No. 114	Low Energy Tests of Conservation Laws in Particle Physics (Blacksburg, VA, 1983)	84-71157	0-88318-313-7
No. 115	High Energy Transients in Astrophysics (Santa Cruz, CA, 1983)	84-71205	0-88318-314-5
No. 116	Problems in Unification and Supergravity (La Jolla Institute, 1983)	84-71246	0-88318-315-3
No. 117	Polarized Proton Ion Sources (TRIUMF, Vancouver, 1983)	84-71235	0-88318-316-1
No. 118	Free Electron Generation of Extreme Ultraviolet Coherent Radiation (Brookhaven/OSA, 1983)	84-71539	0-88318-317-X
No. 119	Laser Techniques in the Extreme Ultraviolet (OSA, Boulder, CO, 1984)	84-72128	0-88318-318-8
No. 120	Optical Effects in Amorphous Semiconductors (Snowbird, UT, 1984)	84-72419	0-88318-319-6
No. 121	High Energy e^+e^- Interactions (Vanderbilt, 1984)	84-72632	0-88318-320-X

No. 122	The Physics of VLSI (Xerox, Palo Alto, CA, 1984)	84-72729	0-88318-321-8
No. 123	Intersections Between Particle and Nuclear Physics (Steamboat Springs, CO, 1984)	84-72790	0-88318-322-6
No. 124	Neutron-Nucleus Collisions: A Probe of Nuclear Structure (Burr Oak State Park, 1984)	84-73216	0-88318-323-4
No. 125	Capture Gamma-Ray Spectroscopy and Related Topics – 1984 (Int'l Symposium, Knoxville, TN)	84-73303	0-88318-324-2
No. 126	Solar Neutrinos and Neutrino Astronomy (Homestake, 1984)	84-63143	0-88318-325-0
No. 127	Physics of High Energy Particle Accelerators (BNL/SUNY Summer School, 1983)	85-70057	0-88318-326-9
No. 128	Nuclear Physics with Stored, Cooled Beams (McCormick's Creek State Park, IN, 1984)	85-71167	0-88318-327-7
No. 129	Radiofrequency Plasma Heating (Sixth Topical Conference) (Callaway Gardens, GA, 1985)	85-48027	0-88318-328-5
No. 130	Laser Acceleration of Particles (Malibu, CA, 1985)	85-48028	0-88318-329-3
No. 131	Workshop on Polarized ^{3}He Beams and Targets (Princeton, NJ, 1984)	85-48026	0-88318-330-7
No. 132	Hadron Spectroscopy – 1985 (International Conference, Univ. of Maryland)	85-72537	0-88318-331-5
No. 133	Hadronic Probes and Nuclear Interactions (Arizona State University, 1985)	85-72638	0-88318-332-3
No. 134	The State of High Energy Physics (BNL/SUNY Summer School, 1983)	85-73170	0-88318-333-1
No. 135	Energy Sources: Conservation and Renewables (APS, Washington, DC, 1985)	85-73019	0-88318-334-X
No. 136	Atomic Theory Workshop on Relativistic and QED Effects in Heavy Atoms (Gaithersburg, MD, 1985)	85-73790	0-88318-335-8
No. 137	Polymer-Flow Interaction (La Jolla Institute, 1985)	85-73915	0-88318-336-6
No. 138	Frontiers in Electronic Materials and Processing (Houston, TX, 1985)	86-70108	0-88318-337-4
No. 139	High-Current, High-Brightness, and High-Duty Factor Ion Injectors (La Jolla Institute, 1985)	86-70245	0-88318-338-2
No. 140	Boron-Rich Solids (Albuquerque, NM, 1985)	86-70246	0-88318-339-0
No. 141	Gamma-Ray Bursts (Stanford, CA, 1984)	86-70761	0-88318-340-4

No. 142	Nuclear Structure at High Spin, Excitation, and Momentum Transfer (Indiana University, 1985)	86-70837	0-88318-341-2
No. 143	Mexican School of Particles and Fields (Oaxtepec, México, 1984)	86-81187	0-88318-342-0
No. 144	Magnetospheric Phenomena in Astrophysics (Los Alamos, NM, 1984)	86-71149	0-88318-343-9
No. 145	Polarized Beams at SSC & Polarized Antiprotons (Ann Arbor, MI & Bodega Bay, CA, 1985)	86-71343	0-88318-344-7
No. 146	Advances in Laser Science—I (Dallas, TX, 1985)	86-71536	0-88318-345-5
No. 147	Short Wavelength Coherent Radiation: Generation and Applications (Monterey, CA, 1986)	86-71674	0-88318-346-3
No. 148	Space Colonization: Technology and The Liberal Arts (Geneva, NY, 1985)	86-71675	0-88318-347-1
No. 149	Physics and Chemistry of Protective Coatings (Universal City, CA, 1985)	86-72019	0-88318-348-X
No. 150	Intersections Between Particle and Nuclear Physics (Lake Louise, Canada, 1986)	86-72018	0-88318-349-8
No. 151	Neural Networks for Computing (Snowbird, UT, 1986)	86-72481	0-88318-351-X
No. 152	Heavy Ion Inertial Fusion (Washington, DC, 1986)	86-73185	0-88318-352-8
No. 153	Physics of Particle Accelerators (SLAC Summer School, 1985) (Fermilab Summer School, 1984)	87-70103	0-88318-353-6
No. 154	Physics and Chemistry of Porous Media—II (Ridgefield, CT, 1986)	83-73640	0-88318-354-4
No. 155	The Galactic Center: Proceedings of the Symposium Honoring C. H. Townes (Berkeley, CA, 1986)	86-73186	0-88318-355-2
No. 156	Advanced Accelerator Concepts (Madison, WI, 1986)	87-70635	0-88318-358-0
No. 157	Stability of Amorphous Silicon Alloy Materials and Devices (Palo Alto, CA, 1987)	87-70990	0-88318-359-9
No. 158	Production and Neutralization of Negative Ions and Beams (Brookhaven, NY, 1986)	87-71695	0-88318-358-7
No. 159	Applications of Radio-Frequency Power to Plasma: Seventh Topical Conference (Kissimmee, FL, 1987)	87-71812	0-88318-359-5
No. 160	Advances in Laser Science—II (Seattle, WA, 1986)	87-71962	0-88318-360-9

No. 161	Electron Scattering in Nuclear and Particle Science: In Commemoration of the 35th Anniversary of the Lyman-Hanson-Scott Experiment (Urbana, IL, 1986)	87-72403	0-88318-361-7
No. 162	Few-Body Systems and Multiparticle Dynamics (Crystal City, VA, 1987)	87-72594	0-88318-362-5
No. 163	Pion–Nucleus Physics: Future Directions and New Facilities at LAMPF (Los Alamos, NM, 1987)	87-72961	0-88318-363-3
No. 164	Nuclei Far from Stability: Fifth International Conference (Rosseau Lake, ON, 1987)	87-73214	0-88318-364-1
No. 165	Thin Film Processing and Characterization of High-Temperature Superconductors (Anaheim, CA, 1987)	87-73420	0-88318-365-X
No. 166	Photovoltaic Safety (Denver, CO, 1988)	88-42854	0-88318-366-8
No. 167	Deposition and Growth: Limits for Microelectronics (Anaheim, CA, 1987)	88-71432	0-88318-367-6
No. 168	Atomic Processes in Plasmas (Santa Fe, NM, 1987)	88-71273	0-88318-368-4
No. 169	Modern Physics in America: A Michelson-Morley Centennial Symposium (Cleveland, OH, 1987)	88-71348	0-88318-369-2
No. 170	Nuclear Spectroscopy of Astrophysical Sources (Washington, DC, 1987)	88-71625	0-88318-370-6
No. 171	Vacuum Design of Advanced and Compact Synchrotron Light Sources (Upton, NY, 1988)	88-71824	0-88318-371-4
No. 172	Advances in Laser Science—III: Proceedings of the International Laser Science Conference (Atlantic City, NJ, 1987)	88-71879	0-88318-372-2
No. 173	Cooperative Networks in Physics Education (Oaxtepec, Mexico, 1987)	88-72091	0-88318-373-0
No. 174	Radio Wave Scattering in the Interstellar Medium (San Diego, CA, 1988)	88-72092	0-88318-374-9
No. 175	Non-neutral Plasma Physics (Washington, DC, 1988)	88-72275	0-88318-375-7
No. 176	Intersections Between Particle and Nuclear Physics (Third International Conference) (Rockport, ME, 1988)	88-62535	0-88318-376-5
No. 177	Linear Accelerator and Beam Optics Codes (La Jolla, CA, 1988)	88-46074	0-88318-377-3
No. 178	Nuclear Arms Technologies in the 1990s (Washington, DC, 1988)	88-83262	0-88318-378-1

No. 179	The Michelson Era in American Science: 1870–1930 (Cleveland, OH, 1987)	88-83369	0-88318-379-X
No. 180	Frontiers in Science: International Symposium (Urbana, IL, 1987)	88-83526	0-88318-380-3
No. 181	Muon-Catalyzed Fusion (Sanibel Island, FL, 1988)	88-83636	0-88318-381-1
No. 182	High T_c Superconducting Thin Films, Devices, and Applications (Atlanta, GA, 1988)	88-03947	0-88318-382-X
No. 183	Cosmic Abundances of Matter (Minneapolis, MN, 1988)	89-80147	0-88318-383-8
No. 184	Physics of Particle Accelerators (Ithaca, NY, 1988)	89-83575	0-88318-384-6
No. 185	Glueballs, Hybrids, and Exotic Hadrons (Upton, NY, 1988)	89-83513	0-88318-385-4
No. 186	High-Energy Radiation Background in Space (Sanibel Island, FL, 1987)	89-83833	0-88318-386-2
No. 187	High-Energy Spin Physics (Minneapolis, MN, 1988)	89-83948	0-88318-387-0
No. 188	International Symposium on Electron Beam Ion Sources and their Applications (Upton, NY, 1988)	89-84343	0-88318-388-9
No. 189	Relativistic, Quantum Electrodynamic, and Weak Interaction Effects in Atoms (Santa Barbara, CA, 1988)	89-84431	0-88318-389-7
No. 190	Radio-frequency Power in Plasmas (Irvine, CA, 1989)	89-45805	0-88318-397-8
No. 191	Advances in Laser Science—IV (Atlanta, GA, 1988)	89-85595	0-88318-391-9
No. 192	Vacuum Mechatronics (First International Workshop) (Santa Barbara, CA, 1989)	89-45905	0-88318-394-3
No. 193	Advanced Accelerator Concepts (Lake Arrowhead, CA, 1989)	89-45914	0-88318-393-5
No. 194	Quantum Fluids and Solids—1989 (Gainesville, FL, 1989)	89-81079	0-88318-395-1
No. 195	Dense Z-Pinches (Laguna Beach, CA, 1989)	89-46212	0-88318-396-X
No. 196	Heavy Quark Physics (Ithaca, NY, 1989)	89-81583	0-88318-644-6
No. 197	Drops and Bubbles (Monterey, CA, 1988)	89-46360	0-88318-392-7
No. 198	Astrophysics in Antarctica (Newark, DE, 1989)	89-46421	0-88318-398-6

No. 199	Surface Conditioning of Vacuum Systems (Los Angeles, CA, 1989)	89-82542	0-88318-756-6
No. 200	High T_c Superconducting Thin Films: Processing, Characterization, and Applications (Boston, MA, 1989)	90-80006	0-88318-759-0
No. 201	QED Structure Functions (Ann Arbor, MI, 1989)	90-80229	0-88318-671-3
No. 202	NASA Workshop on Physics From a Lunar Base (Stanford, CA, 1989)	90-55073	0-88318-646-2
No. 203	Particle Astrophysics: The NASA Cosmic Ray Program for the 1990s and Beyond (Greenbelt, MD, 1989)	90-55077	0-88318-763-9
No. 204	Aspects of Electron-Molecule Scattering and Photoionization (New Haven, CT, 1989)	90-55175	0-88318-764-7
No. 205	The Physics of Electronic and Atomic Collisions (XVI International Conference) (New York, NY, 1989)	90-53183	0-88318-390-0
No. 206	Atomic Processes in Plasmas (Gaithersburg, MD, 1989)	90-55265	0-88318-769-8
No. 207	Astrophysics from the Moon (Annapolis, MD, 1990)	90-55582	0-88318-770-1
No. 208	Current Topics in Shock Waves (Bethlehem, PA, 1989)	90-55617	0-88318-776-0
No. 209	Computing for High Luminosity and High Intensity Facilities (Santa Fe, NM, 1990)	90-55634	0-88318-786-8
No. 210	Production and Neutralization of Negative Ions and Beams (Brookhaven, NY, 1990)	90-55316	0-88318-786-8
No. 211	High-Energy Astrophysics in the 21st Century (Taos, NM, 1989)	90-55644	0-88318-803-1
No. 212	Accelerator Instrumentation (Brookhaven, NY, 1989)	90-55838	0-88318-645-4
No. 213	Frontiers in Condensed Matter Theory (New York, NY, 1989)	90-6421	0-88318-771-X 0-88318-772-8 (pbk.)
No. 214	Beam Dynamics Issues of High-Luminosity Asymmetric Collider Rings (Berkeley, CA, 1990)	90-55857	0-88318-767-1
No. 215	X-Ray and Inner-Shell Processes (Knoxville, TN, 1990)	90-84700	0-88318-790-6
No. 216	Spectral Line Shapes, Vol. 6 (Austin, TX, 1990)	90-06278	0-88318-791-4

No. 217	Space Nuclear Power Systems (Albuquerque, NM, 1991)	90-56220	0-88318-838-4
No. 218	Positron Beams for Solids and Surfaces (London, Canada, 1990)	90-56407	0-88318-842-2
No. 219	Superconductivity and Its Applications (Buffalo, NY, 1990)	91-55020	0-88318-835-X
No. 220	High Energy Gamma-Ray Astronomy (Ann Arbor, MI, 1990)	91-70876	0-88318-812-0
No. 221	Particle Production Near Threshold (Nashville, IN, 1990)	91-55134	0-88318-829-5
No. 222	After the First Three Minutes (College Park, MD, 1990)	91-55214	0-88318-828-7
No. 223	Polarized Collider Workshop (University Park, PA, 1990)	91-71303	0-88318-826-0
No. 224	LAMPF Workshop on (π, K) Physics (Los Alamos, NM, 1990)	91-71304	0-88318-825-2
No. 225	Half Collision Resonance Phenomena in Molecules (Caracas, Venezuela, 1990)	91-55210	0-88318-840-6
No. 226	The Living Cell in Four Dimensions (Gif sur Yvette, France, 1990)	91-55209	0-88318-794-9
No. 227	Advanced Processing and Characterization Technologies (Clearwater, FL, 1991)	91-55194	0-88318-910-0
No. 228	Anomalous Nuclear Effects in Deuterium/Solid Systems (Provo, UT, 1990)	91-55245	0-88318-833-3
No. 229	Accelerator Instrumentation (Batavia, IL, 1990)	91-55347	0-88318-832-1
No. 230	Nonlinear Dynamics and Particle Acceleration (Tsukuba, Japan, 1990)	91-55348	0-88318-824-4
No. 231	Boron-Rich Solids (Albuquerque, NM, 1990)	91-53024	0-88318-793-4
No. 232	Gamma-Ray Line Astrophysics (Paris-Saclay, France, 1990)	91-55492	0-88318-875-9
No. 233	Atomic Physics 12 (Ann Arbor, MI, 1990)	91-55595	088318-811-2
No. 234	Amorphous Silicon Materials and Solar Cells (Denver, CO, 1991)	91-55575	088318-831-7
No. 235	Physics and Chemistry of MCT and Novel IR Detector Materials (San Francisco, CA, 1990)	91-55493	0-88318-931-3

No.	Title		
No. 236	Vacuum Design of Synchrotron Light Sources (Argonne, IL, 1990)	91-55527	0-88318-873-2
No. 237	Kent M. Terwilliger Memorial Symposium (Ann Arbor, MI, 1989)	91-55576	0-88318-788-4
No. 238	Capture Gamma-Ray Spectroscopy (Pacific Grove, CA, 1990)	91-57923	0-88318-830-9
No. 239	Advances in Biomolecular Simulations (Obernai, France, 1991)	91-58106	0-88318-940-2
No. 240	Joint Soviet-American Workshop on the Physics of Semiconductor Lasers (Leningrad, USSR, 1991)	91-58537	0-88318-936-4
No. 241	Scanned Probe Microscopy (Santa Barbara, CA, 1991)	91-76758	0-88318-816-3
No. 242	Strong, Weak, and Electromagnetic Interactions in Nuclei, Atoms, and Astrophysics: A Workshop in Honor of Stewart D. Bloom's Retirement (Livermore, CA, 1991)	91-76876	0-88318-943-7
No. 243	Intersections Between Particle and Nuclear Physics (Tucson, AZ, 1991)	91-77580	0-88318-950-X
No. 244	Radio Frequency Power in Plasmas (Charleston, SC, 1991)	91-77853	0-88318-937-2
No. 245	Basic Space Science (Bangalore, India, 1991)	91-78379	0-88318-951-8
No. 246	Space Nuclear Power Systems (Albuquerque, NM, 1992)	91-58793	1-56396-027-3 1-56396-026-5 (pbk.)
No. 247	Global Warming: Physics and Facts (Washington, DC, 1991)	91-78423	0-88318-932-1
No. 248	Computer-Aided Statistical Physics (Taipei, Taiwan, 1991)	91-78378	0-88318-942-9
No. 249	The Physics of Particle Accelerators (Upton, NY, 1989, 1990)	92-52843	0-88318-789-2
No. 250	Towards a Unified Picture of Nuclear Dynamics (Nikko, Japan, 1991)	92-70143	0-88318-951-8
No. 251	Superconductivity and its Applications (Buffalo, NY, 1991)	92-52726	1-56396-016-8
No. 252	Accelerator Instrumentation (Newport News, VA, 1991)	92-70356	0-88318-934-8

No. 253	High-Brightness Beams for Advanced Accelerator Applications (College Park, MD, 1991)	92-52705	0-88318-947-X
No. 254	Testing the AGN Paradigm (College Park, MD, 1991)	92-52780	1-56396-009-5
No. 255	Advanced Beam Dynamics Workshop on Effects of Errors in Accelerators, Their Diagnosis and Corrections (Corpus Christi, TX, 1991)	92-52842	1-56396-006-0
No. 256	Slow Dynamics in Condensed Matter (Fukuoka, Japan, 1991)	92-53120	0-88318-938-0
No. 257	Atomic Processes in Plasmas (Portland, ME, 1991)	91-08105	0-88318-939-9
No. 258	Synchrotron Radiation and Dynamic Phenomena (Grenoble, France, 1991)	92-53790	1-56396-008-7
No. 259	Future Directions in Nuclear Physics with 4π Gamma Detection Systems of the New Generation (Strasbourg, France, 1991)	92-53222	0-88318-952-6
No. 260	Computational Quantum Physics (Nashville, TN, 1991)	92-71777	0-88318-933-X
No. 261	Rare and Exclusive B&K Decays and Novel Flavor Factories (Santa Monica, CA, 1991)	92-71873	1-56396-055-9
No. 262	Molecular Electronics—Science and Technology (St. Thomas, Virgin Islands, 1991)	92-72210	1-56396-041-9
No. 263	Stress-Induced Phenomena in Metallization: First International Workshop (Ithaca, NY, 1991)	92-72292	1-56396-082-6
No. 264	Particle Acceleration in Cosmic Plasmas (Newark, DE, 1991)	92-73316	0-88318-948-8
No. 265	Gamma-Ray Bursts (Huntsville, AL, 1991)	92-73456	1-56396-018-4
No. 266	Group Theory in Physics (Cocoyoc, Morelos, Mexico, 1991)	92-73457	1-56396-101-6
No. 267	Electromechanical Coupling of the Solar Atmosphere (Capri, Italy, 1991)	92-82717	1-56396-110-5